BUILD-IT-YOURSELF

RODALE PLANS

Home Soyfood Equipment

For Production and Use of High-Protein, Low-Calorie Tofu, Tempeh, and Soymilk

edited by
Ray Wolf

Rodale Plans
Rodale Press
33 East Minor Street
Emmaus, PA 18049

Projects and plans designed and published by Rodale Press have been researched and tested by its in-house facilities. However, due to the variability of all local conditions, construction materials, and personal skills, etc., Rodale Press assumes no responsibility for any injuries suffered or damages or other losses incurred during or as a result of the construction of these designs. We recommend that before major construction is undertaken the plans are first reviewed and approved by a knowledgeable local architect or builder for feasibility and safety as well as compliance with all local and other legal and code requirements. All instructions and plans should be carefully studied and clearly understood before beginning any construction.

Rodale Plans
Rodale Press
33 East Minor Street
Emmaus, PA 18049

Printed in the United States of America on recycled paper containing a high percentage of de-inked fiber.

Library of Congress Cataloging in Publication Data

Wolf, Ray.
Home soyfood equipment.

(Rodale plans)
1. Soybean processing. 2. Soybean as food. I. Title.
II. Series.
TP438.S6W64 664′.64 81-8551
ISBN 0-87857-361-5 paperback AACR2
2 4 6 8 10 9 7 5 3 1 paperback

Technical Illustrator
Frank Rohrbach

Writers
Jim Hoffman
Carol Keough

Project Designers
Philip Gehret
Jim Eldon
Keith Marks
Ray Wolf

Test Kitchen
Linda C. Gilbert
Susan Hercek
Debra Deis

Testing and Evaluation
Harry E. Wohlbach

Copy Editor
Felicia D. Knerr

Photographer
John Hamel

Cover Photograph
Margaret Smyser

Book Layout
Merole Berger

Table of Contents

Section I Introduction

This book was like an amoeba; every day it seemed to grow. I have the feeling that, if it had not stopped when it did, it would have become an encyclopedia, then a series or maybe a collection—perhaps even the world's first soyfood library.

From my original idea of a simple book that told how to build equipment to produce soyfoods at home conveniently, we've arrived at what can only be called a hybrid book. This is not just a book that tells you how to build something—it's much more. It's a cookbook, and it's also a book that looks at the ecology of what we eat.

More than any other Rodale book I can think of, this has truly been a company-wide effort. First, our staff of designers, working under Jim Eldon, began working with our test kitchen staff to design the equipment. Linda Gilbert, with her kitchen staff of Debra Deis and Susan Hercek, tested numerous pieces of prototype equipment. They kept coming back, asking for more refinements. At one point we felt it was hopeless; a piece of equipment that was easily built and did all they asked of it just couldn't be developed. Working from the original designs developed by Keith Marks, Phil Gehret designed the presses and tofu box presented here. He kept refining them to please the test kitchen staff, and when he finally hit on it, they were as happy as he was. Then our designers turned their attention to the tempeh incubator, while the kitchen crew began working on ways to use the presses and testing early models of incubators.

Gerald Hock built our very first tempeh incubator. We felt it was just a prototype and began trying to improve it. But improvements proved to be elusive. The more we worked, the tougher it got, until we reached the point where our "refined" designs barely worked. Day after day we had more failures than successes. Finally, in an effort to regroup, I went back to Jerry's first design and slightly modified that. Success! A few more modifications by Phil, and we had an incubator that constantly produced top-quality tempeh.

Through all this, Linda's kitchen staff worked to find, not only ways to make the foods, but totally new ways to use the foods as well. I can't give this group enough credit for their inventive work. We turned them loose, and they have done things with tofu, soy beverages, and tempeh I did not think possible. Their techniques and recipes far surpass my greatest wishes. I'm sure you'll be happy with them. It was reassuring to hear the comments at our daily taste-testing go from "less than palatable," to "I'd serve this at home," and, best of all, "do you have any more left?"

Once we had designs, Jim Hoffman began the task of telling you how to build the units, and Frank Rohrbach began the illustrations for the book, while working with Jim on the methods of construction. The Product Testing Department assigned Harry Wohlbach to evaluate the designs. While this was going on, Kim Greenawalt began typing the ever-growing manuscript between learning how to make the foods and writing up that information.

We brought Carol Keough, the Organic Living editor from *Organic Gardening* magazine, in to rewrite the test kitchen's report; this now appears as the first two chapters. Carol not only made the information enjoyable and readable, she did it quickly and on time. I owe her a big thank you.

Most of all, I'd like to thank Rodale Press for letting this book become a reality. Of the more than ten Rodale books I've been involved with, in many ways I see this as the best and most important. We've taken an entire grouping of foods, found new and better ways to make and use these foods, and put it all between two covers.

You, the reader, can't begin to appreciate the amount of work that went into this book. I know all the people who worked so hard to make it possible, and they all have my gratitude. For me, this book is better than I had ever hoped it would be. I know how valuable it can be, and I hope you use it and share it with others.

Ray Wolf

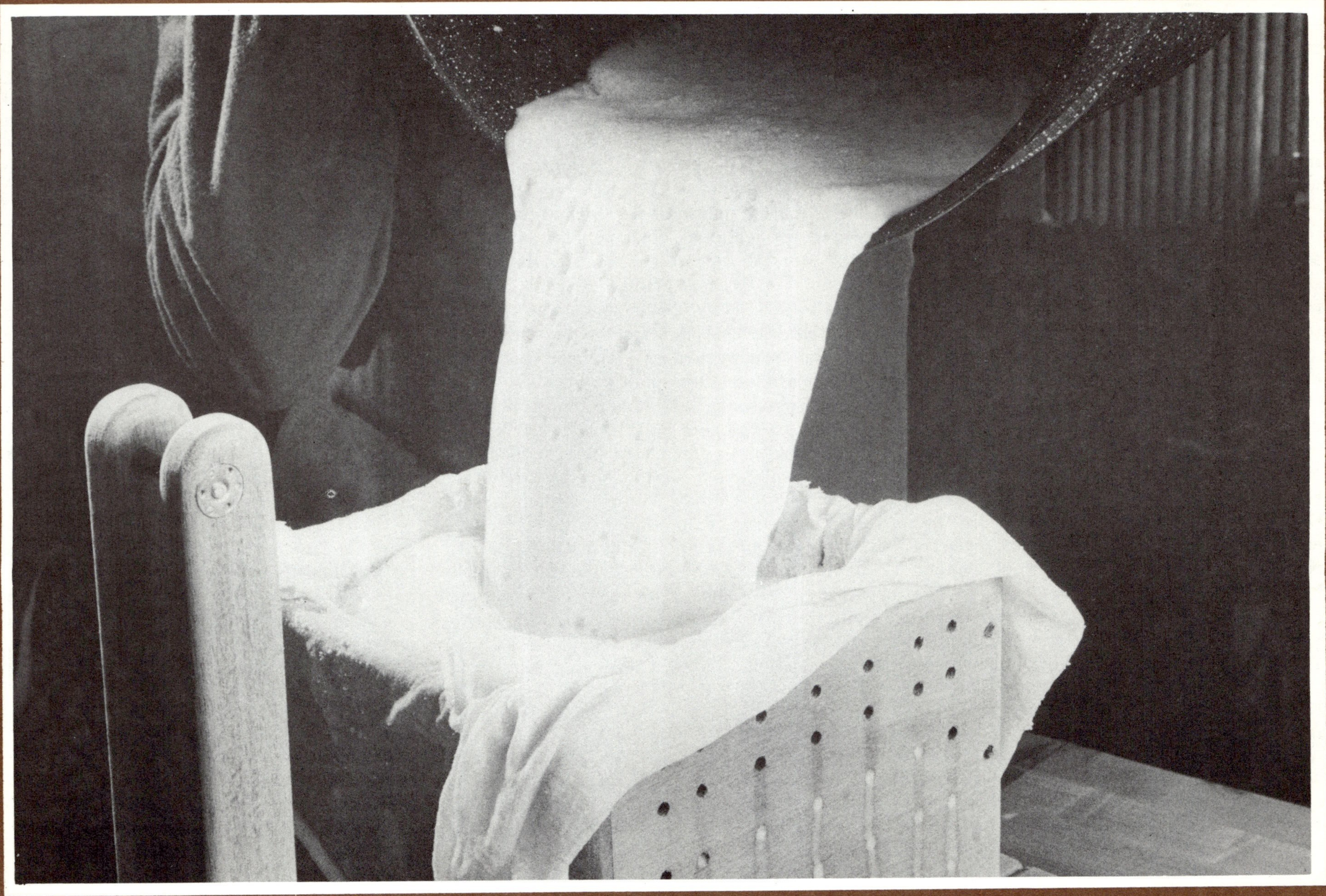

1 What Soyfoods Are

Years ago, cartoonist Al Capp created an intriguing character for his comic strip "L'il Abner." Called the Schmoo, it lived just to please. If someone was hungry, the Schmoo turned into food. If that person wanted steak, the Schmoo was steak. If it was a baby who wanted milk, the Schmoo tasted like milk. Even better, there was no end to the Schmoos, because as soon as one bunch was gobbled up, there was another bunch marching over the line from the next cartoon panel.

What marvelous fantasy! America fell in love with the Schmoo and celebrated Al Capp for his wisdom. We all talked about what a wonderful world it would be if something like the Schmoo really existed.

Well, folks, there really is a food that's a lot like the Schmoo. While it won't cuddle up in your lap, tickle you with its whiskers, and kiss you the way Capp's Schmoo would, it will become just about any kind of food you want. Want a hamburger? It's a hamburger. Milk? Sure enough, it's milk. Or cheese, or custard, or cereal, or crackers, or pudding, or even a souffle!

What's more, this marvelous food is loaded with protein, vitamins, and minerals. And, like the Schmoo, it's your good friend because it won't make you fat or clog your arteries with cholesterol. Even better, it's cheap, cheap, cheap. You can buy it in stores or make your own.

Of course, we're talking about soyfood—a food relatively new to this country, but a staple in the Orient for thousands of years.

NECESSITY, THE MOTHER OF SOYFOOD

Given the large population of East Asia and the limited amount of farmland, it's not surprising that the Chinese, Japanese, Indonesians, and other East Asians have always used their precious land to grow nutritious crops rather than for animal pastures. After all, a cow requires about 20 pounds of protein-rich feed to produce 1 pound of protein. With that ratio, they simply could not even begin to feed themselves on a meat-based diet. The Orientals deduced that if the feed is protein rich, why not just skip the animal and eat the feed?

Because plain soybeans don't taste all that hot, that's why. Not only don't beans taste very good by themselves, our bodies can't use many of the nutrients in the bean without some form of processing. In America, we process beans and other grains by feeding them to livestock. And what a processor livestock is! In go beans; out come steaks and roasts. Unfortunately, although livestock does create a very tasty end product, it isn't very cost effective.

The Orientals looked for different ways to process beans—ways that were much more efficient and still yielded a desirable product. What finally evolved were two basic types of foods: tofu (and its sister product, soymilk) and tempeh. If you want to get 1 pound of

Protein Processing Yields

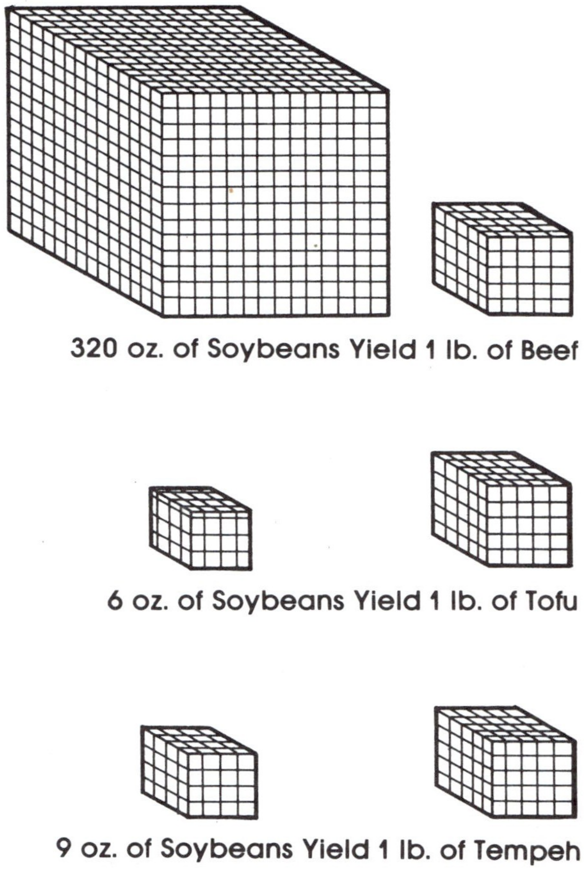

Chart 1-1—This shows the differing efficiencies of several methods of processing soybeans. Feeding beans to cattle is the least efficient way to produce an edible protein, while tofu is the most efficient.

food from soybeans, chart 1-1 shows several processing options. As you can see, the cow is the least efficient, by far. Thus, you don't see many grazing animals in the Orient; they've gone the way of the 8 MPG, 1958 American Buick—to the scrapyard. The future belongs to the efficient.

And so, over thousands of years, a cuisine has evolved that depends mainly on grains (rice) and legumes (soybeans). Of course, people could not live very long eating plain rice and beans three times a day, every day, without going crazy. And sure enough, their cuisine developed to include a wide variety of delicious foods with different flavors and different textures, all based on the simple but efficient grain and legume. The range of soyfoods includes beans, cheese, paste, milk, fermented meat substitutes, and others.

Tradition holds that much of the early development of soyfoods was done by Whi Nain Tze, of Huai-nan, China, about 165 B.C. He was a scholar, ruler, politican, and philosopher who created a way to extract "milk" from soybeans. And, by curdling the milk with seawater, he soon produced a type of cheese called tofu (toe-foo). With beans, milk, and cheese as the cornerstone ingredients, Oriental cuisine was off and cooking.

Soybeans were eventually introduced into the United States at the beginning of the nineteenth century. However, for many years they were grown only as a rotation crop and for forage. We felt they were good for the land and our animals but not for us. Today, this country produces two-thirds of the world's soybeans. They are used for cooking oil, margarine, paint, varnish, soap, glue, plastics, lubricants, and, most of all, livestock feed. In addition, we export thousands of tons of beans every year, making them an important commodity in international trade and a big plus on the credit side of our balance of payments ledger. However, only recently, beginning with the beef shortage of the early 1970s, have Americans tried eating soybeans in earnest.

As a nation of meat eaters, we initially tried to overpower the bean and to use soyfoods as meat substitutes. Food technologists battled with the bean, first creating protein concentrates from it. Then, by defatting, spinning, extruding, compressing, and otherwise manipulating the basic bean, it became possible to make imitation ham, beef, sausage, and chicken. As you might have tasted, imitation meats left a lot to be desired. As far as public acceptance, they went over like the legendary lead balloon.

So poor were some of those initial soyfoods that the bean quickly got a bad reputation. It's a shame, really, because soybeans have much to offer us. In addition to a fine natural flavor and versatility in cooking, they provide excellent nutrition and economy. The problem is, America's food industry tried to overpower the bean and lost. The Orientals succeeded in using the bean by working with it, not against it. Over time they have learned the bean's strengths and weaknesses and have developed a series of foods that emphasize the bean's particular strengths.

Perhaps the key to success is understanding that soyfoods are not meant to stand alone; they are ingredients—ingredients in a varied cuisine. Sort of a base food, something to be used along with something else. Soyfoods are the nutritional foundation on which the largest single segment of the earth's population has based its diet. Soyfoods need to be understood before being used.

This book introduces three basic soy foods: tofu, soy beverages, and tempeh (tem-pay). Tofu is cheese made from soybeans, and the clone of the Schmoo. It has a gentle flavor and a soft, moist texture. Added to a casserole or a stir-fry dish, it assumes the flavors of the ingredients around it. Soy beverages are made from the liquid extract of the bean—creamy, rich, and sometimes tasting mildly of the bean from which they came. Tempeh is a fermented bean food from Indonesia. It has an unassuming flavor that some people describe as nutty, with a soft, "nuggety" texture.

Not only are soyfoods easy on the land while greatly increasing the food-producing capacity of the land, they are extremely economical. At this writing, a pound of soybeans costs an average of 54 cents. One cup of whole beans will make six to seven cups of soymilk or 1 pound of tofu. Two cups will provide 1¼ to 1½ pounds of tempeh.

Thus, if you are feeding a family of four a casserole that contains 1 pound of hamburger at $1.90 a pound, you will save $1.36 by substituting tofu for hamburger. Aha, you may say, the savings aren't worth sacrificing the family's nutrition for. Perhaps the best thing about soyfoods is that they require no nutritional sacrifices. In fact, adding soyfoods to your diet will improve your overall nutrition, not harm it.

Traditionally, when dollars get tight, families suffer cutbacks in their diets. These cuts often take the form of substituting potatoes, pasta, or grains for meat, creating precipitous drops in protein consumption while expanding waistlines. These foods, used to replace meat, are filling but have far less protein. They put on pounds and do little to fuel our bodies.

However, a switch to soyfoods should end up improving your family's nutrition. They are a good protein source but are very low in fat and calories. They offer goodly amounts of minerals and most vitamins, with the exception of vitamin C, and they contain no cholesterol.

Chart 1-2 compares the nutritional makeup of soyfoods to several more traditional foods. You'll see that the lowly bean holds up very well when compared to the far more costly meat products and exceedingly well when compared to the more traditional meat replacers. That is a key part of understanding and appreciating soyfoods. Don't compare soyfoods to ground round; compare them to rice, potatoes, pasta, and other grain products. That's where they really shine. Use soyfoods to reduce the amount of meat you use, not necessarily to eliminate it.

Besides economy and good nutrition, there are still more reasons to use and enjoy soyfoods. They are low in calories, thus perfect for dieters. They are very easy to digest, so easy, in fact, that babies and even invalids

Nutrient Content of Various Foods

Food	Water (%)	Calories	Protein (g)	Fat (g)	Total Carbohydrates (g)	Fiber (g)	Ash (g)	Calcium (mg)	Phosphorus (mg)	Iron (mg)	Sodium (mg)	Potassium (mg)	Vitamin A (IU)	Thiamine (mg)	Riboflavin (mg)	Niacin (mg)	Vitamin C (mg)
Tofu	84.8	72	7.8	4.2	2.4	0.1	0.8	128	126	1.9	7	42	...	0.06	0.03	0.1	...
Tempeh, Soy	60.4	157	19.5	7.5	9.9	1.4	1.3	142	240	5.0	...	...	...	0.28	0.65	2.52	...
Soybean Milk	92.4	33	3.4	1.5	2.2	...	0.5	21	48	.8	...	...	40	0.08	0.03	0.2	...
Hamburger, Cooked	54.2	286	24.2	20.3	...	...	1.3	11	194	3.2	47	450	40	0.09	0.21	5.4	...
Ham, Roasted	45.5	374	23.0	30.6	...	...	0.9	10	236	3.0	...	...	...	0.51	0.23	4.6	...
Chicken Drumstick, Fried	55.0	235	32.6	10.2	1.0	...	1.2	15	236	2.3	...	...	140	0.07	0.40	7.1	...
Cow's Milk	87.4	65	3.5	3.5	4.9	...	0.7	118	93	Tr.	50	144	140	0.03	0.17	0.1	1
American Cheese	39	398	25.0	32.2	2.1	...	3.7	750	478	1.0	700	82	1,310	0.03	0.46	0.1	...
Potatoes, Baked in Skin	75.1	93	2.6	0.1	12.1	0.6	1.1	9	65	0.7	4	503	Tr.	0.10	0.04	1.7	20
Rice, White Enriched, Cooked	72.6	109	2.0	0.1	24.2	0.1	1.1	10	28	0.9	374	28	...	0.11	...	1.0	...
Spaghetti, White Enriched, Cooked	72.0	111	3.4	0.4	23.0	0.1	1.2	8	50	0.9	1	61	...	0.14	0.08	1.1	...

Chart 1-2—This compares the nutrients in soyfoods to those in meats and traditional meat replacers, per 100 grams of food. The critical point is the ratio of calories to protein. With soyfoods, you need to consume far fewer calories in order to receive your needed amount of protein.

can maintain a high-level diet. They are on the low end of the food chain, so they have not accumulated the assortment of industrial chemicals like PCB and PBB that is found in meat and fish. Organically grown, soybeans provide food that is free from the pesticide and herbicide residues that are found on many commercially grown fruits and vegetables.

Finally, it's good for your soul to switch to soyfoods. One-quarter of the people of the world are hungry. As Americans, we represent only 6 percent of the world's population, but we eat 30 percent of its meat. Land which could provide food in abundance is turned over to inefficient cattle. While one individual or one family cannot end the suffering of millions by changing their diet, it is a step in the right direction. And, as the ancient Chinese proverb states, "Even the longest journey begins with a single step."

One last encouragement: As the world's population gets ever-larger, and the amount of food-producing land continues to decline, the traditional, meat-based American diet begins to look more and more like an endangered species. Why not begin switching now, while you can do it at your own pace? You'll learn not only to accept these new foods, but to enjoy them as well.

THE SOYFOODS

Before going any further, we should explain what you are about to read. The remainder of this chapter is divided into three parts. Each of the three basic soyfoods will be explained. When we explain a food, we'll tell you not only how to make it, but also how to use the equipment you will be building.

If soyfoods do have one drawback, it is the time it has always taken and the trouble it has always been to prepare the foods. In fact, these problems were the initial spark behind this book. Our test kitchen wanted to work with soyfoods but found the time and trouble to make them just too much to overcome. They asked our designers to develop some equipment to help them out, and the whole project was off and running. The equipment you will be reading about and building is not available anywhere else. We have designed and perfected it to be built at home with only the simplest of tools and most basic of skills. No one produces it commercially. The techniques of making the foods that go with the equipment are also new and are pegged directly to this equipment.

And now, let's go on to examine these three intriguing foods closer.

TOFU

Tofu is a soft cheese made from soymilk. It has a smooth texture, fragile composition, and a custardlike consistency. It is high in protein and low in fat and calories and has absolutely no cholesterol or salt. In the Orient, it is considered the perfect food.

And no wonder! You can alter the consistency of tofu by adjusting the amount of pressure used to form it into a cake, leaving it soft or solid. Once prepared, it can be used as cakes, cubes, or slices. It can be mashed, pressed, or squeezed. It can be pureed or beaten into a crumbly mass.

Unlike other soft cheeses, which keep only three or four days before they begin to lose their appeal, tofu can be stored in the refrigerator for as long as ten days. And should it turn, you can refresh it by parboiling. If you want to keep tofu longer, it's possible to freeze it.

Its versatile form and adaptable flavor make it a food for all seasons—and all seasonings. Tofu's high protein content makes it an excellent meat substitute or extender. Because tofu is very bland, you simply add the spices or seasonings you would normally use for meat, and voilà! Tofu can taste like seafood, sausage, Italian meatballs, whatever! And a chicken casserole without fowl will taste remarkably like chicken if poultry seasoning or stock is used.

Tofu also works as a beneficial addition to high-fat, high-salt, high-cost cheeses. By combining the two products, you achieve the richness of flavor associated with the cheese, but you get more cheese and better nutrition. You also can use tofu with dips, puddings, sauces, and a myriad of other dishes.

In addition to tofu, the pressed soybeans also yield okara and whey. Okara is the pulp that remains after the soymilk has been pressed from the beans. Of all soyfoods, it is fibrous okara that blends most easily into a recipe without making its presence noticeable. Okara is a high-fiber food, and more palatable than the ever-popular wheat bran. When cooked in milk, it evokes childhood breakfasts of farina. Fresh, it looks like ground coconut.

Either fresh or dried and ground into a fine powder, okara can add nutrition and fiber to almost any recipe. Drinks, soups, cereals, puddings, cookies, meat loaves, and burgers all benefit from the addition of a little okara.

Whey is the clear liquid pressed from the curds when making tofu. It contains some protein, most of the water-soluble B vitamins, and some carbohydrates. It not only can be used to enrich soups, stews, and other dishes with a liquid component but, if substituted for whatever liquid is required in a recipe, whey can add nutrition to any food that comes out of your kitchen. Additionally, whey has a lathering property. Traditionally, it has been used to clean the tofu equipment after all the work is done.

Generally, the method for making tofu or traditional soymilk includes soaking the beans, grinding them into a slurry, heating and pressing the slurry to produce a soymilk, adding a solidifier to curdle the soymilk, and, finally, pressing the curds into tofu.

Each step has a very distinct purpose. The beans are soaked overnight to hydrate and tenderize them. They must absorb water before grinding. During grinding most of the protein is brought into solution. You need the protein in liquid form so it can be removed from the nonprotein part of the bean, the fiber. The ground beans are then added to boiling water. This further tenderizes the bean, but, most important, it destroys the trypsin inhibitor. To simplify a lot of chemistry, the trypsin inhibitor prevents your body from utilizing all the protein in the beans. Cooking enables your body to use all the protein. When the water and beans reach boiling they will suddenly foam up, and you will be ready to express the milk.

When you express the milk, you are separating the liquid portion from the solid portion. The liquid portion carries soluble protein and vitamins and minerals, while the solid portion carries the fiber, some protein, and vitamins and minerals, but most of the nutrients are now in the liquid. Then you return the soymilk to the stove and slowly heat it to boiling. A solidifier is added to change the pH of the milk. This causes the protein to solidify into small curds floating in whey. The curds are ladled out of the whey, returned to the box, and pressed together into a piece of tofu, expressing additional whey. After the tofu is pressed, it is soaked in cold water to further firm it up. It is then stored in a covered container of fresh water to prevent it from drying and getting hard and rubbery.

Before detailing exactly how to make tofu (pages 22-23), let's look at the equipment you will need. We've designed three pieces of equipment; you will need two of them, a tofu box and a press. The box, illustration 1-1, has four sides with a removable top and bottom. All four sides and the bottom have holes drilled in them, to allow soymilk and whey to drain from the box. The box sits on one of the two presses, and the inside of the box is lined with cheesecloth. You pour soymilk into the box and the cheesecloth filters out the okara, allowing most of the pure soymilk to drain out of the box. Then you put the top on the box and use the press to squeeze out any additional milk. Traditionally, this step was done by wringing the hot okara by hand, a some-

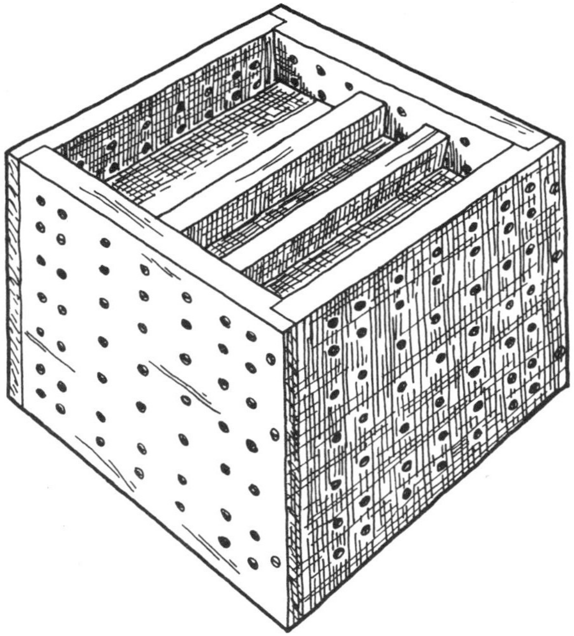

Illustration 1-1—The tofu box is used in conjunction with one of the presses. It is adjustable for 1, 3, or 5 pounds of tofu. It is also used for expressing soymilk.

times painful, always inefficient method.

The soymilk is then reheated and a solidifier is added to it. This causes the protein in the milk to coagulate. The curds of protein are then carefully ladled into the box, which is again lined with cheesecloth. This time, you work very gently to prevent the curds from breaking apart. You wrap the cloth over the curds and reapply the top to the box. The plunger then comes down and applies very slight pressure to expel the whey and form the loose curds into a solid block—a piece of cheese. The amount of pressure you apply will dictate whether you have firm or soft tofu.

A unique feature of the box is that it is adjustable. Two baffles can be installed in the box at two different positions, giving the box the versatility needed to be the perfect size for three different amounts of tofu. Using the baffles, the box is properly sized for 1- or 3-pound batches of tofu, while without the baffles, the box is sized for 5 pounds of tofu. There also are different-size pressure plates. Use the proper pressure plate for the adjusted baffle position.

The instructions on pages 22-23 should be closely followed whenever you make tofu, until you are sure you know the process well. Because making tofu, like making bread, is more a craft than a process, go slowly and carefully until you develop a feel for it.

The tofu box is designed for use with a press. We have designed two presses, one to mount on top of a canning kettle, the other to be used on a counter top. Read about both presses to decide which one you will build. We have found the pot-mount press to be our favorite. We make large batches of tofu and have canning kettles available. If we did not have a kettle or much floor space in the kitchens, we would prefer the table-mount press.

POT-MOUNT PRESS

The pot-mount press, detailed in illustration 1-2, is designed to work on top of a 7-quart water bath canning kettle measuring about 9 inches high and 14 inches in diameter. Be sure to soak the press and box in warm water for two to three hours and air-dry it well before its first use, to "cleanse" the wood.

Set the kettle on the floor, and unfold the press frame. Place the bottom frame rails on top of the kettle so that the notches on the bottom fit inside the rim of the kettle. These notches prevent the frame from slipping off the kettle when you are using it.

The arms of the frame should be in a vertical position. Place the bottom of the tofu box on the frame rails so its two feet run in the opposite direction from the frame rails. The feet should fit between the frame rails perfectly. Put the tofu box over its bottom, so it fits evenly and is seated directly on top of the frame's rails.

With the box and bottom in position on the press, you have to assemble the lever and plunger. The plunger is removable, but should be attached to the lever when the unit is built. The plunger should fit into the box, and the notch on the end of the lever fits around the copper pipe at the top of the arms of the press. The bottom of the plunger fits between the guides on the pressure plate.

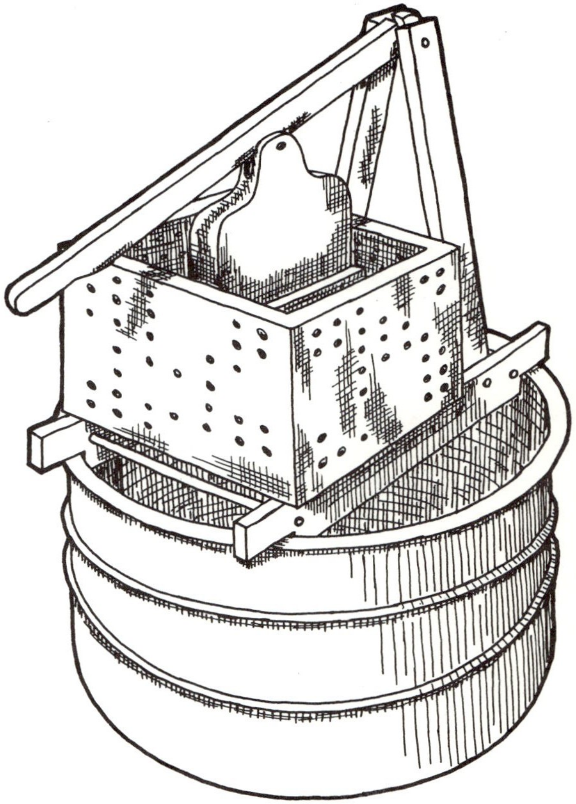

Illustration 1-2—The pot-mount press sits atop a standard water bath canning kettle. The press can be used to press tofu or to express soymilk.

The press allows you to apply two kinds of pressure. By leaning down on the lever, great pressure can be applied, as needed for expressing soymilk. However, for forming the tofu cake, very little pressure is needed, but it must be constant. For this, we have included a small cup hook on the end of the lever to hang a weight. By experimenting, you'll find what weight produces the type of tofu you like. We have found that cast-iron frying pans make ideal weights.

That should be all you need to know to use the pot-mount press. It really is quite simple. It is a platform with arms that mounts on a canning kettle. The tofu box sits on the platform, slurry or curds are poured into the box, and the soymilk or whey drains into the kettle below.

TABLE-MOUNT PRESS

The table-mount press shown in illustration 1-3, as its name implies, was designed to be used on a table or a bench. Before using it for the first time, soak it in warm water for several hours, then let it air-dry.

To set the press up, simply unfold it so that the frame is standing on all four legs, and the arms are in a vertical position, pulled out of their inlaid position in the frame. The press needs a metal tray, large enough for the tofu box to sit in. A cookie sheet is ideal. The tray should be modified when you build the press. It should have sides about 1 inch high and holes in one corner. This tray will catch the milk and whey coming out of the box, and the holes will let the liquid drain into a bowl placed under the press.

Place the tray on the press so its long side is flush against the arms. Then put the bottom of the box on the tray. Set the tofu box on the bottom so it fits flush. Slide the box as close as possible to the arms. With the box in position, you have to attach the lever and plunger. This is done the same for both presses.

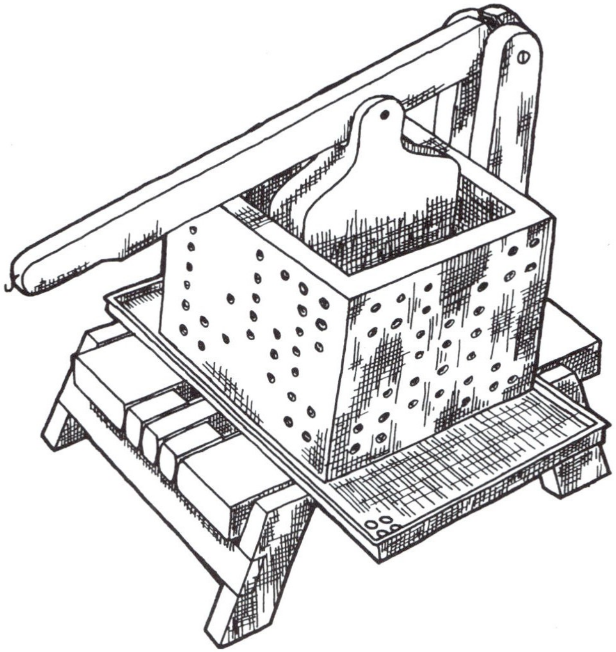

Illustration 1-3—The table-mount press sits on a table or counter top. The press can be used to press tofu or to express soymilk.

To collect soymilk or whey you will need a pot or dish no greater than 4 inches in height. Put this dish directly underneath the holes in the tray.

That should be all you need to know to use the table-mount press. It really is quite simple. It is a platform with arms that sits on a table or bench. The tofu box sits on a tray on the platform, slurry or curds are poured into the box, and the soymilk or whey drains into the tray and into a bowl below the tray.

CLEANING THE PRESSES

Because soymilk is sticky and hard to remove, it is important to clean the presses as soon as you have finished using them. Do not allow the soymilk to dry on the wood.

To clean either of the presses and the tofu box, you will need a very small bottle brush that will fit into the drain holes of the tofu box and a good, coarse scrub brush.

Begin cleaning by rinsing all parts of the press and box with hot water. Using the brush, scrub out all the holes in the tofu box bottom, sides, and baffles. Use only hot water or hot whey to clean the presses. Do not use detergent. If you use whey as a cleaning agent, be sure to rinse all parts of the press with clear water when finished. If you find the press is hard to clean, make a solution of ¼ cup baking soda and ½ gallon water to use with the scrub brush.

Once all parts of the press have been cleaned satisfactorily, place them on a clean cloth, and allow them to air-dry for several hours. Do not put the box or presses into a dishwasher. The washing action will not clean them well enough, and the drying cycle is not good for the wood.

MAKING TOFU

When making soyfood products, it is important to begin with good soybeans. Look for the yellow variety rather than the green—or choose the lightest beans of those available to you. Always use whole soybeans, because they provide a finished product that is higher in protein than one made with cracked beans.

Once you have selected your beans, store them in a cool, dry place. Dryness is a very important factor, because beans stored in a damp spot can develop dangerous molds or simply rot before you can use them. Properly stored beans can last about a year before deterioration begins.

Use untreated (not chlorinated) water for tofu: bacteria-free water from a private spring or well or bottled water. The vast improvement in flavor is worth almost any amount of trouble required to provide pure water.

Finally, select the type of solidifier you plan to use for curdling the soymilk. In the ancient Orient, pure ocean water was used as the solidifier (nigari). However, that commodity no longer exists. Instead, you can use the following: epsom salts (our preferred choice), Japanese nigari, calcium chloride, distilled white vinegar, rice vinegar, or lemon juice.

Each coagulant contributes its own quality to the ultimate flavor of the tofu, so that subtle variations in taste are possible because of the type of solidifier you choose. Calcium chloride offers a subtly sweet flavor. Tofu solidified with epsom salts is mild and soft. For a tarter, firmer tofu, choose lemon juice or vinegar.

Consult chart 1-3 to find the amount of solidifier you will need for each cup of dry beans. Experiment with all the solidifiers until you find the one you like best. Calcium chloride and epsom salts are available at any drug store. The vinegar and lemon juice are found in any supermarket. Try to use the freshly squeezed juice rather than the reconstituted bottled lemon juice. Nigari can be found in some health food stores or in Oriental grocery shops.

To make 1 pound of firm tofu you will need one cup of whole yellow soybeans, the chosen solidifier, and 3½ quarts of pure water.

Whether you use the pot-mount press or the table-mount press, you will need the following equipment: a large wooden spoon; a ladle to remove the curds from the whey; measuring spoons and measuring cups; a colander for draining the beans; a 2-quart bowl for soaking one cup of dry beans and later for holding the bean slurry; a 4-quart cooking pot with a heavy bottom; a blender to grind the beans and water into soy slurry; and three lengths of cotton cheesecloth, each measuring 32 inches by 22 inches. Two pieces will be used to line the straining box when you are pressing the soymilk, and later the third piece will be used for pressing tofu. You will need quality cotton cheesecloth, the kind used in cheese making, normally found in fabric stores. Do not use the packaged dustcloths sold as cheesecloth, because they are not strong enough to stand up to the pressing procedure.

Solidifier Amounts
(per cup of dried beans)

Solidifier	Amount
Epsom Salts (magnesium sulfate)	1 1/4 teaspoons
Nigari	1 1/4 teaspoons
Calcium Chloride	1 1/4 teaspoons
Distilled White Vinegar or Rice Vinegar	2 tablespoons
Lemon Juice (fresh)	8 teaspoons

Chart 1-3—The amount of various solidifiers needed differs. Use the indicated amount of the solidifier you select for each cup of beans.

PRESEASONING TOFU

In addition to all the procedures described for making tofu, and detailed on pages 22-23, it is also possible to preseason the soymilk so that the end product will be flavored tofu.

Our test kitchen has found that the addition of fresh scallions, parsley, and basil leaves—all finely chopped—gives tofu a piquant flavor and an attractive appearance. After pressing, the tofu cake will be flecked with green.

The amount of seasoning varies greatly, depending on which kind you use. Apply your knowledge of cooking to create interesting flavors. And remember, it is better to start with too few spices than too many.

If the herbs and spices you choose require a bit of cooking time to blend their flavors, add them earlier in the tofu-making process. We have found that adding herbs just after pressing the milk allows time for them to mellow as the milk is being warmed for the addition of the coagulant.

If you like to experiment, you can create endless varieties of tofu. Try preseasoning 1 pound of tofu by adding a tablespoon of grated onion, leek, or garlic to the soymilk immediately after pressing. If the results please you, go creative, experiment.

That should be about all you need to know to get involved in tofu making. Pages 22-23 are designed to be used whenever you make tofu. Follow the step-by-step directions on these pages, and you should have no trouble making successful tofu. Now, let's look at some of the beverages you can make from soybeans.

SOY BEVERAGES

The first time most people ever hear of soymilk is in a pediatrician's office. Normally the exchange goes like this one, related by Carol Keough.

"For several months I had a visiting schedule that coincided with another young mother's. You couldn't help notice that her baby was cranky, thin, and usually suffering with a crusty, red rash. One day, she left the examination room on the verge of tears. 'The doctor thinks Kevin is allergic to his formula. He wants me to put the baby on soymilk,' she said in a thin, anxious voice. I sympathized. Soymilk sounded terrible—even worse than the goat milk they had previously tried."

"But when we met again for our next monthly visit, Kevin's cheeks were rosy, his body slightly round, and his disposition remarkably changed. For the first time, I saw him smile. We admired his two new front teeth. 'And you know what?' his mother asked, 'He *likes* soymilk!'"

In the 20 years that have passed since Baby Kevin popped his two new teeth, the world has heard plenty about soymilk, not all of it good. At Rodale Press, we've tried some fabulous concoctions in our company dining room, Fitness House. We've enjoyed piña colada shakes, soy "ice cream," creamed soups, and the more traditional honeyed soymilk. And even the most skeptical among us admit that soymilk, in all its varied forms, is good for you.

However, although good for you, many people find the taste of soymilk objectionable. "It may be all right for a baby who can't drink anything else, but not for me," is the standard retort. Bearing this in mind, our test kitchen set out to develop a better soymilk.

Now, there are two types of soy beverages, traditional soymilk and our newly developed beverage base. Technically, beverage base is just good-tasting soymilk, but for clarity, we will differentiate between the two as we continue.

The traditional soymilk is slightly higher in protein than beverage base, and will make larger quantities of tofu.

Why, then, develop a beverage base? Because in taste tests Rodale employees have unanimously favored the flavor of the beverage base. It needs only small amounts of flavoring to prepare a good-tasting drink. On the other hand, the traditional soymilk has beany flavor notes and is often called painty in flavor. It needs to be flavored heavily to mask its undesirable taste.

Thus, you should make traditional soymilk to make tofu, and beverage base for a drinking liquid. Chart 1-4 compares the nutritional content of the two drinks. The higher protein content of soymilk gives a higher yield when making tofu. The better-tasting beverage base will yield less tofu. As you'll see in the next chapter, there are certain times when you may want to make soymilk for something other than tofu. Beverage base will not clabber; thus you cannot use it for some of the recipes we have developed. For those recipes under clabbered soymilk, you have to make soymilk, not beverage base.

We give step-by-step procedures for making both soymilk and beverage base. To make soymilk, follow the instructions on pages 22-23 for making tofu, and stop at step 11. To make beverage base, follow the in-

Soymilk vs. Beverage Base

	% in Soymilk	% in Beverage Base
Crude Protein	3.1	2.1
Moisture	93.5	95.6
Fat	1.4	0.9
Ash	0.4	0.3
Crude Fiber	0.2	0.1

Chart 1-4—The nutritional makeup of the beverage base and soymilk are very similar, the main differences being protein content and taste. Beverage base is better for drinking, soymilk for making tofu.

structions on page 24.

Beverage base has a flat, almond color and a bland taste that lends itself well to adaptation, from adding a drizzle of honey or a drop of vanilla to more exotic treatments. Because it does not have the amount of fat that dairy milk has, it does not have the creamy texture of whole milk. It feels somewhat thin on the tongue.

Nutritionally, beverage base contains slightly less protein than dairy milk, is low fat, and contains absolutely no saturated fat. Because its calcium and B-vitamin content is notably lower than that of animal milk, if you are on a diet of beverage base only, supplement your diet with these nutrients.

Covered and in the refrigerator, soymilk will keep for at least 11 or 12 days. Like dairy milk, soymilk will sour after prolonged storage. In this form it can be used as clabbered milk or buttermilk, cup for cup, in recipes calling for those ingredients. Beverage base will not sour; it will just spoil. Like dairy milk, beverage base can be used to make yogurt; it will curdle if it's boiled, and it will stick to the bottom of the saucepan if it's burned. Unlike dairy milk, it will not whip, because it is so low in fat. It can be frozen or canned, enjoyed fresh or made into sauces, soups, ice milk, or shakes. It can substitute for dairy milk in any recipe, and you use exactly the amount called for.

In working with soybeans, it is important to understand two factors inherent in the bean that can affect the nutrition and flavor of your beverages. The first is an antinutritional factor called the trypsin inhibitor, which we've already talked about. In very simple terms, soybeans contain an element that prevents your body from breaking down some of the protein, so it passes through without becoming available to your body.

Fortunately, food scientists have discovered that this antinutritional factor can be inactivated by boiling the beans. Moreover, they have found that by adding bicarbonate of soda to the blanching water, making the pH of the water more alkaline, the trypsin inhibitor is destroyed even more rapidly. As luck would have it, the addition of bicarbonate of soda also tenderizes the soybeans.

The second factor, which can affect the flavor of soyfoods, is the lipoxidase enzyme. Often people will complain that soy products taste beany or painty. Scientists have found that these off-flavors come from an enzyme present in the bean. The enzyme that causes all this trouble is, of course, the lipoxidase enzyme. It, too, can be routed by heat. We use this feature to great advantage when preparing beverage base.

The process of making beverage base is really quite simple. You can use either press, but you need the tofu box. The first step is to bring water, with baking soda added to it, to a boil. Then add dry beans to the boiling water for overnight soaking.

It is important that the water be boiling when the beans are added. If there were any cracked beans, the beany flavor would be formed as soon as they got wet. By adding dry beans to boiling water, the enzyme that creates the beany flavor is immediately inactivated, and no beany flavor forms. The baking soda in the soaking water tenderizes the beans and helps to destroy the trypsin inhibitor.

Another feature of the process is that the baking soda increases the pH of the water. When you add soybeans to hot water, their protein solubility decreases, and there will be less protein in the milk. However, the increased pH counteracts this trend, and you end up with a soymilk of good protein content.

The next morning, pour the soaked beans into a colander and rinse the beans well. Bring more water to a boil, and put the boiling water and beans in a blender and grind them. You will have to process several batches because of the blender's limited capacity. Again, it is important that the beans be in very hot, almost boiling water when they are broken to prevent the beany flavor from forming.

The blended slurry is then added to the press, and the soymilk is expressed. Only in this case, you are expressing beverage base—soymilk with a vastly improved taste. Your beverage base is then heated in a double boiler for 20 minutes, and it is ready to be used, as detailed in chapter 2.

To make beverage base, follow the step-by-step instructions on page 24.

TEMPEH

A traditional staple in the diet of Indonesians, tempeh is a whole food, rich in fiber, B vitamins, minerals, and protein. Tempeh also has a reputation for providing vitamin B_{12}, an exclusive nutrient for many vegetarians. However, researchers at Cornell University have found that this is a product of a bacteria (*Klebsiella*), which may not be present in every batch of tempeh.

It is ideal as a meat substitute, because it has a full-bodied, chewy texture. It can be shaped like hamburger patties and served on a bun, sculpted into cutlets, sautéed in butter, deep fried, combined in casseroles, or crumbled over the top of dishes to increase their protein content dramatically. Recipes for all these dishes and more appear in chapter 2. It is a versatile, tasty food that deserves a place in the American kitchen. Tempeh can be made from soybeans or other legumes, like chick-peas, fava beans, okara (a by-product of soymilk), or even peanuts. Each type of tempeh has its own unique flavor.

There are two methods of making tempeh—one is the traditional method, which requires soaking the beans overnight, hulling them (a tedious, time-consuming task), cooking them, then inoculating them with a starter and incubating them for 24 hours in a specially built container. This method is reliable and produces tempeh of excellent quality; however, it is very time-consuming. For example, to serve tempeh for a Saturday night dinner, you would have to start preparing it on the previous Thursday evening.

The Rodale Test Kitchen has developed a new method that requires only 45 minutes of preparation time, plus the 24-hour incubation period. The trick to this method is to substitute roasted, unsalted soy halves (also called soy nuts) for whole, dry soybeans. This substitution eliminates both the overnight soak and the time-consuming hulling process. It also reduces the cooking period from 45 minutes to only 10 minutes.

Chart 1-5 compares the two methods of making tempeh for nutritional content. The methods produce an almost-identical end product. Complete instructions for preparing short-method tempeh are on page 25.

The only drawback to the short method is that you must buy commercially prepared roasted soy halves. These are normally available in almost any health food store and many supermarkets or nut shops. Be sure you get soy halves that are roasted, but not salted. At this writing, 1 pound of tempeh produced by the long method will only cost 32 cents, while a pound by the short method costs 41 cents, not counting the cost of starter. We think the almost 4 hours of preparation time you save is well worth the extra dime it costs to use the short method.

Like making good wine or cheese, making tempeh is a craft. The creation of tempeh, like cheese, is dependent on a beneficial mold. As you make tempeh more and more frequently, you will learn the habits and peculiarities of the mold and soon you will find yourself working with confidence and ease.

Long-Method Tempeh vs. Short-Method Tempeh

	% Using Long Method	% Using Short Method
Crude Protein	22.1	21.9
Moisture	57.5	56.4
Fat	9.4	8.0
Ash	1.1	1.6
Crude Fiber	2.5	1.7

Chart 1-5—Long-method tempeh compared to short-method tempeh. Note that the nutrient content is almost identical; the difference is in the preparation time.

You begin by cooking the soy nuts for 10 minutes in water to which vinegar has been added. The addition of a small amount of vinegar will lower the pH of the soybeans, making them slightly acid, thus creating a better medium for the tempeh mold. The water you use should not be municipal water, because chlorine also restricts mold growth. If you can manage it, use well water or spring water. If you can afford it, use bottled spring water.

After the beans have been cooked for about 10 minutes, drain them, then let them dry and cool. While this sounds like a very simple step, it is crucial to making good tempeh. If the beans go into the incubator too wet or too dry, your tempeh probably will not be successful. Allow 20 minutes for drying and cooling, but don't go off to watch the "Late Show," because if you leave the beans for more than about 30 minutes, they will be too dry.

A second important factor is cleanliness. The world is full of ubiquitous mold spores and bacteria. What you want to do is create a specific environment that favors the tempeh mold, but knocks out as many of the others as possible. Therefore, it is important to clean the incubator thoroughly after each use, to use only freshly washed utensils, pots, and pans, and to have hands so clean you could pass for a surgeon.

Once the beans have cooled and dried, you will inoculate them with tempeh starter, a mold called *Rhizopus oligosporus*. It can be ordered in two different forms. A freeze-dried, powdered starter can be ordered from Farm Foods, 156 Drakes Lane, Summertown, TN 38483. This starter is quite easy to use. Just put it in an old salt shaker and sprinkle it over the soybeans. At this writing, enough powdered starter to treat 3 pounds of beans costs $2, and enough for 25 pounds costs $10.

A starter comprised of mold spores held in agar-agar and contained in a petri dish is available from Gem Cultures, 30301 Sherwood Road, Fort Bragg, CA 95437. While this starter is more reliable, it is a bit more difficult to use, because you will have to break up the culture and work it through the soybeans carefully and uniformly. Each of these cultures costs $1.50 and will process 2 to 4 pounds of soybeans.

While the directions on page 24 for making tempeh attempt to give an exact measure for the inoculum, the amount you add to the beans will vary with the strength of starter. Fortunately, you don't have to be a chemist to figure it out—the suppliers of the starter will tell you the strength and how much to add to the soybeans in their directions.

Once the beans have been inoculated, they will be placed in uniform layers, no more than ½ to ¾ inch deep, in two baking pans. You have your choice of pans: muffin tins, tart tins, 9-inch by 9-inch metal or glass baking pans, or any other suitably sized pan. Be sure the containers DO NOT have high sides, which could impede the airflow inside the incubator and interfere with proper mold formation. It is best to build your incubator first, then buy pans to fit it.

You want to grow a mycelium (mold) that is velvety white and present throughout the entire mass of beans, as shown in photo 1-1. This mold will cover the tempeh until separate pieces of bean are united into a firm cake. The mycelium will digest both the fat and protein in the bean, turning the hard pieces into chewy nuggets. Enzymes in the mold break down the protein into amino acids, making that nutrient more available to the body. Moreover, the mold itself produces B vitamins as a by-product.

When the tempeh is finished, the mold should completely cover the beans from top to bottom and side to side. Uneven growth of the mold indicates trouble. For instance, it is possible to have a good coat of mold over the top of the beans while the underneath may be

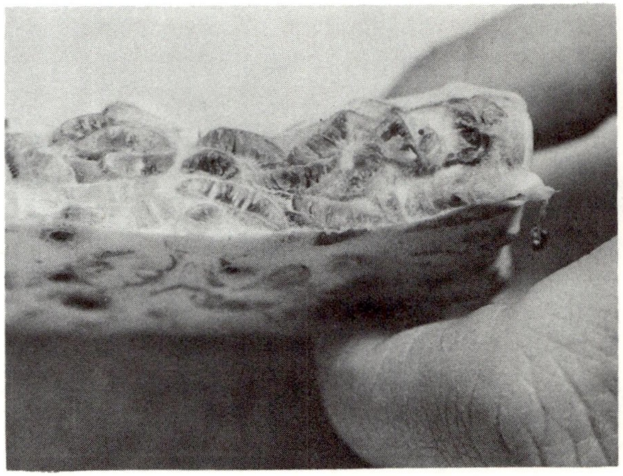

Photo 1-1—Good tempeh looks like this. The mold should be throughout all the beans.

yellowish, runny, and very, very stinky when you lift up the cake.

Therefore, it is important to do whatever you can to give the mold a chance to develop throughout the beans. If you use a metal pan, which we recommend, you can facilitate the growth of the mold by perforating the bottom of the container with 1/16-inch holes, ½ to ¾ inch apart, in a uniform grid pattern. In that way, the spores can get a foothold on the underside of the tempeh cake as well as on the top surface.

Using a thermometer, test the temperature inside the incubator before putting the beans inside. A range of 88° to 95°F is fine, but we recommend aiming for exactly 92°F. With that temperature, the incubator's heat can fluctuate a few degrees up or down without destroying the mold.

With the beans in the pans, place the pans in the incubator and settle down. Patience is required for the next 24 hours.

If you peek, you won't see much happening to the tempeh for the first 10 hours or so. However, after 10 or 12 hours, a few fine white strands may appear. Mold has been present all along, but it hasn't been visible.

During the incubation period, moisture will be formed in the incubator. Some of this moisture may condense on the sides and top of the incubator. There is nothing you can do to prevent it. We've tried. After about 12 hours, you can wipe out the incubator with a clean cloth or sponge. You don't have to dry the incubator, but it may help your tempeh to grow better. Contaminated tempeh will produce more moisture than uncontaminated tempeh, so keep an eye on the moisture. Tempeh will not grow if the relative humidity in the incubator gets much over 85 percent. Wiping off the condensed moisture will help reduce the humidity.

You will know your tempeh is "done" when the entire cake is uniformly covered in mold. It will look clean, white, pure—like freshly fallen snow. It will smell faintly yeasty, somewhat like freshly picked mushrooms. If it smells bad, it's bad. Photo 1-3 shows the difference between good tempeh and bad tempeh.

If the mold is predominantly white but has a few gray patches, it is fine. It may have incubated just a little bit too long so that some black spores (like fruit) have formed, but it is perfectly safe to eat.

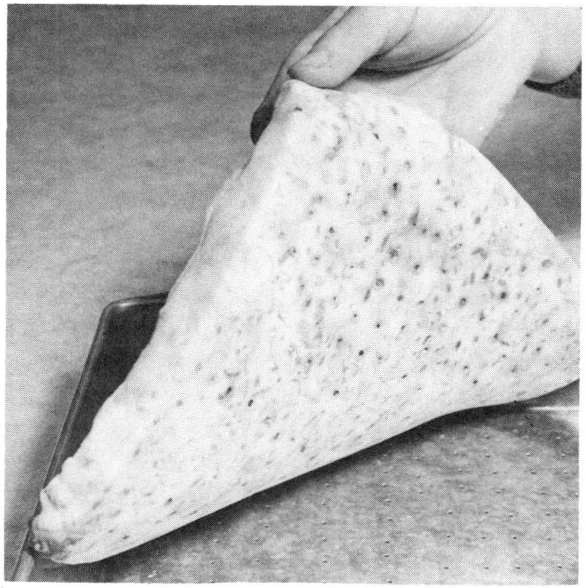

Photo 1-2—When finished, the tempeh will form a soft cake with flecks of soybeans present.

Danger signs of bad tempeh include yellow slime, yellow mold, a strong ammonia smell, or other offenders. If the smell is moldy but not unpleasant, the tempeh is OK. The color must be white or light gray.

After the tempeh has completed incubation, it can be cooked immediately, refrigerated, or frozen. Be sure to clean the incubator after each use. Wipe the interior surfaces with a damp cloth, using a mild detergent, if necessary. Leave the box open to air out and dry thoroughly.

If a batch of contaminated tempeh has grown in the incubator, wipe it out immediately with a solution of ammonia and water. Wipe out the box again with a clean, damp cloth, then allow it to dry thoroughly.

Do not spray the incubator with disinfectant or submerge it in water. Do clean it with ammonia and water every few weeks.

As you'll see in chapter 6, the incubator is nothing more than an insulated plywood box with a reflective interior and a light tray. The lights are controlled by a thermostat to keep the heat steady.

Incubator designs were tried and refined until we developed an incubator that (1) is easy to build; (2) works perfectly, with little attention required from the cook; (3) uses very little energy; and (4) is attractive to look at and will fit on a kitchen counter under a wall-hung cabinet. Our finished design is shown in illustration 1-4.

Photo 1-3—The pan on the left holds good tempeh, while the tempeh in the other pan has been contaminated.

Inside, two wire cake-cooling racks serve as shelves for the trays of incubating tempeh. After a great deal of testing, it was found that putting the trays on the same level gave better results than placing one over the other. This arrangement helps to overcome any problems with heat stratification within the incubator. The box is top-loading, making it easy to build and easy to load and unload.

The incubator is easy to use. A steady temperature of about 92°F is needed for the proper mycelium formation, so it is important to preheat the incubator and check the temperature with a thermometer before placing your tempeh inside. If the temperature is not within a degree or two of 92°F, you can raise or lower it by adjusting the temperature regulator. For more heat, you simply turn the adjusting screw until the lights come on; for less heat, turn it until they go off. Wait a few minutes for the incubator to regulate itself, and then recheck the temperature. Continue checking until the 92°F is achieved.

When the beans have been inoculated, just place them on the cake racks inside, making sure they do not touch the sides of the incubator. If they do touch, the containers will not receive even heat, and the mold will not spread properly through your tempeh cake. Replace the incubator top so no heat can escape.

The exact length of incubation time varies with the temperature. The lower the heat, the longer growth takes. However, the incubation period should never exceed 30 hours.

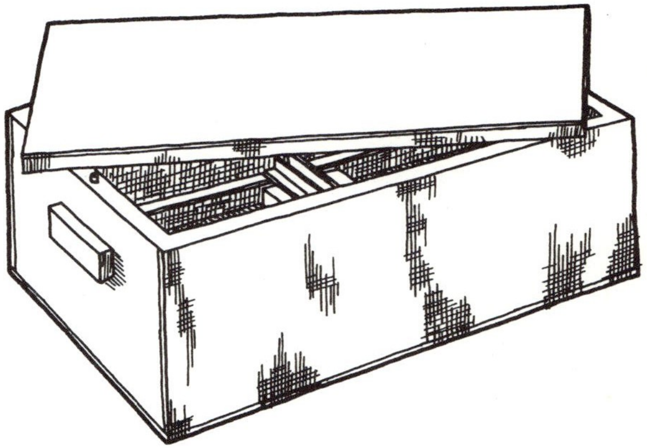

Illustration 1-4—The tempeh incubator is a simple-to-build insulated box with a thermostatically controlled heat source.

MAKING STARTER

If you find yourself preparing tempeh more and more often, and you probably will, you'll find that making your own starter is a worthwhile and money-saving skill. A technique for making tempeh starter using rice was developed by Drs. Hwa L. Wang, E. W. Swain, and C. W. Hesseltine at the Northern Regional Research Center, Agricultural Research Service, of the United States Department of Agriculture, and adapted by the Rodale Test Kitchen.

You will need a pressure cooker; pint mason jars; and screw bands and milk filters, which will replace the lids of the jars. If milk filters are not available, you can make your own by putting ⅛ to ¼ inch of cotton between two layers of gauze. These filters keep the dust and bacteria out of the jar, but allow the mold to breathe.

Here are the simple steps for making starter. Remember to use clean utensils.

1. Place ¼ cup brown rice in a pint mason jar, add ¼ cup spring water, mix well, cover with the milk filter in place of the lid, and put on the screw band.

2. Let the jar stand at room temperature for 1 hour, and shake it every 10 or 15 minutes so the rice will take up the water uniformly.

3. Cook the rice in the jar in a pressure cooker for 20 minutes at 15 pounds of pressure. The water in the pressure cooker should

be 3 inches deep. You need enough water to generate steam, but not so much that it will get inside the jars.

4. Shake the jar to break up the rice lumps, cool, and add ⅛ teaspoon of freeze-dried tempeh starter. Put back the filter and screw top, then mix the rice and starter well by shaking the jar.

5. Lay the jar on its side and place it in the incubator at 92°F for four days. At that time, the rice should be covered with black spores. Pulverize the spore-covered rice in a dry, clean blender for 1 or 2 minutes or until it looks like uniform gray granules. Get it as fine as you can. If it is too pasty to grind, freeze it, then grind it. The starter is now ready to use.

6. Put the rest of the starter back into the mason jar, and cover the jar with the filter, the solid lid, and the screw ring. Keep the jar in the freezer, where the starter will remain viable for months.

The strength of this homemade starter may vary. You will have to experiment with it to determine the amount required for successful inoculation. Start with 2 teaspoons per batch. You may need more or less.

That should be all you need to know about making tempeh. Page 25 gives exact processing instructions for making tempeh; follow them closely.

Tropical Tofu Cheesecake

Crust

1 1/4 cups ground almonds
1/4 cup dried unsweetened coconut
3/4 cup whole wheat pastry flour
1/4 cup mild oil
2 tablespoons honey
1/2 teaspoon vanilla

Filling

2 1/2 cups unsweetened pineapple juice
1 cup dried unsweetened coconut
2 tablespoons unflavored gelatin
2 tablespoons water
1 cup chopped dried unsweetened pineapple
1/2 cup chopped dried unsweetened papaya
2 tablespoons honey
2 pounds medium-firm tofu, crumbled (about 4 cups)
6 tablespoons mild oil
2 teaspoons vanilla
2 tablespoons lemon juice
3/4 cup mashed ripe bananas

Garnish (optional)

thinly sliced dried papaya and pineapple
sprigs of mint or blossoms

1. To make crust, combine the almonds, coconut, and flour. In another bowl, combine the oil, honey, and vanilla, then work the wet mixture into the dry with a pastry blender or your hands until well combined. Press into an oiled 9 x 3-inch springform pan, 2 inches up the side. Bake at 400°F 8 to 10 minutes, until lightly toasted. Cool.
2. To make filling, bring juice and coconut to a boil in a saucepan. Reduce heat, cover, and simmer 5 minutes. Pour mixture into a blender. Start at the lowest setting and gradually raise it to the highest. Blend about 30 seconds. Press mixture well through a jelly bag, doubled cheesecloth, or very fine sieve. The coconut can be saved for use in other recipes or for toasting.
3. Stir gelatin into water. Bring coconut-juice mixture back to a boil. Stir in gelatin until dissolved, then add dried fruit and remove from the heat. Stir in honey.
4. Combine tofu, oil, and vanilla. Stir lemon juice into mashed bananas and add this, along with the dried fruit and juice mixture, to the tofu. Stir together well. Blend 1/3 of the mixture at a time, starting on low speed and gradually working up to the highest speed. Blend until very smooth, thick, and creamy. Pour blended portions into another bowl and mix together well.
5. Pour filling into prebaked crust. If you intend to decorate the cheesecake, do it now with the dried fruit. Chill overnight, preferably 24 hours. Remove from pan and further garnish with mint or blossoms.

Yield: 12 servings

AN AWARD WINNER

This Tropical Tofu Cheesecake recipe won second place at the 1981 National Tofu Cheesecake Bake-Off, sponsored by the Soycrafters Association of North America. The competition is the vegetarian alternative to the Betty Crocker Bake-Off.

The cheesecake is high in protein and very low in both saturated fat and sodium, but, like most cheesecakes, fairly high in calories.

The recipe was developed by the Rodale Test Kitchen's product development staff. Susan Hercek was the principal researcher.

Making Tofu

INGREDIENTS

Soybeans—Use whole, yellow soybeans. Be sure the beans have not been treated and are suitable for human consumption. Select lighter colored beans with a light hilium (the circular, dimplelike area on the bean). Different varieties will yield almost the same; **1 cup dry beans yields 1 pound of tofu; 3 cups, 3 pounds; and 5 cups, 5 pounds.**

Solidifier—A wide variety of solidifiers may be used. The most dependable soldifier is magnesium sulfate, common epsom salts. Chart 1-3 (page 14) gives the recommended amount of all solidifiers to use. Use **1¼ teaspoons epsom salts to 1 cup water per pound of tofu.**

Water—This is a surprisingly important ingredient of tofu. The better-tasting the water is, the better your tofu is likely to be. Untreated spring water is best. If using chlorinated or fluoridated water, let it sit in an uncovered container for several days to "exhaust" most of the chemical odor. Tofu requires a lot of water. Plan on using **15 cups water for every pound of tofu.**

EQUIPMENT

Spoon—large, long-handled
Measuring Spoons—a variety of sizes
Measuring Cups—both wet and dry measures
Colander—large enough to drain the beans
Ladle—a hand strainer can be used
Bowl—2-quart capacity per pound of tofu, with cover
Large Pot—6-quart for 1 pound of tofu, canning kettle for 3 or 5 pounds
Small Pot—1-quart for 1 pound, 2-quart for 3 or 5 pounds
Containers—to hold the slurry; 2-quart for 1 pound, 6-quart for 3 pounds, 10-quart for 5 pounds
Blender—capacity of the blender is unimportant
Cheesecloth—fine-weave cotton cheesecloth (obtainable at fabric stores); three pieces, each 32 inches by 22 inches
Weights—Standard cast-iron frying pans are ideal. An 8-inch skillet produces tofu of an average firmness.
Press and Box—either Rodale's pot-mount or table-mount soyfood press and tofu box.

PROCEDURE

1. **Soak beans** at room temperature in a covered container for 8 to 16 hours. Overnight is best. Soak 1 cup of beans in 4 cups of water per pound of tofu.

2. **Drain off excess water** after soaking by gently pouring the beans into a colander.

3. **Boil water,** 5 cups for every pound of tofu. This will take some time; start now, and proceed to the next step. Use a pot with a larger capacity than the amount of water you will be boiling; for 1 pound of tofu use a 6-quart pot; for 3 pounds use a canning kettle or a 12-quart pot; and for 5 pounds use a canning kettle.

4. **Prepare slurry** while the water is coming to a boil. Depending on the size of your blender and the amount of tofu you are making, your procedure will differ. The goal is to blend the soaked beans and water. For 1 pound of tofu, use $2^2/_3$ cups of water; for 3 pounds use 8 cups of water; and for 5 pounds use $13^1/_3$ cups of water. Put some beans and water into the blender, start at a low speed, and increase the speed until the mixture is smooth, $1^1/_2$ to 2 minutes. If you have a large blender and are making a small batch of tofu, you can do this in one step. If you have a small blender or are making a large batch of tofu, it could require many batches. When one batch of slurry is complete, do not add it to the heating water; hold it in a separate container and slurry another batch. All of the slurry should be held in a container and added to the boiling water at once, in the next step.

5. **Set up press** and mount the box on the press. Wet two of the pieces of cheesecloth to line the inside of the box. The cloth should be put into the box from opposite sides with the pieces overlapping each other to cover the bottom and four sides of the box completely. The cheesecloth should overhang the top of the box on all sides.

6. **Heat the slurry** by pouring all of it into the boiling water. It is important that all the slurry be added to the boiling water at one time, as quickly as possible. After adding the slurry, keep the heat on, but turn it down slightly.

7. **Heat rinse water** for the okara (the soybean pulp) in a separate pot. This will be poured through the okara after the soymilk has been expressed. For each pound of tofu, you will need 2 cups of heated water.

8. **Boil the slurry,** stirring frequently to prevent it from scorching and sticking to the pot. Bring the mixture to the boiling point. As soon as the mixture reaches the boiling point, it will quickly foam up. The foaming action can be

quite sudden and dramatic. As soon as it foams up, turn off the heat and remove the pot.

9. **Separate the milk** by pouring the heated slurry into the cheesecloth-lined tofu box. This separates the soymilk from the okara. The soymilk will drain through the cheesecloth quite quickly. Pour enough to fill your container under the press, change containers if need be, and pour more milk, until all the slurry is poured into the box. Then fold the cheesecloth over the mixture remaining in the box and put the pressure plate on top of the folded cheesecloth. Connect the handle and plunger to the press, and push with all your weight to express the remaining milk. If using the pot-mount press, steady it as you push on the handle.

10. **Rinse the okara** with the hot rinse water. Unfold the cheesecloth from the okara, stir the okara, and pour the rinse water over it. Refold the cheesecloth and re-press the okara, to get out the last of the soymilk. When this is completed, remove the okara from the box. It may be used in this form, or dried and stored. Rinse out your cheesecloth.

11. **Stop here for traditional soymilk.** The liquid you have strained from the okara is soymilk, which can be used in many ways. See chapter 2 for details on using traditional soymilk and okara. To make tofu, continue.

12. **Boil the soymilk** by returning it to the stove over a medium-high heat to prevent scorching. Stir frequently so the milk does not stick to the sides or burn. While the milk is heating, proceed to the next step.

13. **Prepare the solidifier** by dissolving it in water. The solidifier we recommend is epsom salts. Add the solidifier to the water, and stir until dissolved. For each pound of tofu, dissolve 1¼ teaspoons in 1 cup of water.

14. **Add one-third of the solidifer** when the soymilk reaches an active boil. Stir it in gently, but well. After stirring it in, turn off the heat, but leave the pot over the burner. Hold the spoon still in the middle of the pot to help stop the turbulence created by stirring.

15. **Add the second third of the solidifier** when all motion has stopped in the pot. Slowly pour the solidifier over the back of the spoon, letting it gently dribble into the soymilk as it runs off the spoon. You must do this gently, and distribute the solidifier over the entire area of the pot. Cover and let stand for 3 minutes.

16. **Add remaining solidifier** by again dribbling it into the milk, using the back of the spoon to break the pouring force of the solidifier. Again, working delicately, make sure the entire surface of the liquid is covered with the splashing liquid. Stir the top ½ inch of liquid very gently after the solidifier is added. Cover the pot and let stand for 6 minutes.

17. **Light, delicate curds** should begin forming in the soymilk after the second batch of solidifier is added. The curds will look somewhat like white scrambled eggs floating in a clear yellowish liquid. If, after adding the last of the solidifier and waiting 6 minutes, you still have trouble getting curds to form, mix a little more solidifier and add it gently to the mix.

18. **Prepare the press** while waiting for the curds to form. Set up the press and use the third piece of cheesecloth. Wet the cheesecloth and form it to the inside of the box. Leave the ends of the cheesecloth overhanging the top of the box.

19. **Remove the curds,** working very gently. Ladle the curds out of the pot and into the cheesecloth in the box. Be very careful not to break the curds when moving them from pot to box. If you have trouble removing the curds, very gently pour the pot of whey through the box to trap all the curds in the cheesecloth. This must be done carefully to prevent breaking the curds.

20. **Press the tofu** once all the curds are in the cheesecloth. Fold the cheesecloth over the curds, put the pressure plate on top of the curds, and set the handle and plunger on the pressure plate. Hang a weight from the hook on the end of the handle. Some whey will drain from the curds while the curds are being formed into the block of tofu. Don't use a lot of pressure on the handle when doing this. Leave the weight on the handle for 15 to 30 minutes to firm up the tofu; the longer the weight is on, the firmer the tofu.

21. **Remove the tofu** by gently moving the box, holding the bottom, to a sink of cold water. Submerge the box and the tofu in the water, then remove the box, and slip the bottom of the box away from the cake of tofu. You should now have a cake of tofu wrapped in cheesecloth soaking in water. Let it soak for a few minutes to cool. Then remove the cheesecloth and store the tofu in a covered container of water in the refrigerator.

22. **Clean all equipment** carefully as explained in chapter 1.

Making Beverage Base

INGREDIENTS

The ingredient amounts that are given will make 6 cups of beverage base; double all amounts for 12 cups.

Soybeans—Use whole, yellow soybeans. Be sure the beans have not been treated and are suitable for human consumption. Select lighter colored beans with a light hilium (the circular, dimplelike area on the bean). Different varieties of beans produce beverage base with different subtleties of flavor, but the yield from different varieties should be almost the same; **1 cup of dry beans yields 6 cups of beverage base.**

Baking Soda—This is a very important ingredient, as it tenderizes the beans, makes the protein more soluble, and helps to prevent the typical beany taste in soymilk. Use standard baking soda (sodium bicarbonate); **6 cups beverage needs 1/4 teaspoon baking soda.**

Water—This is a very important ingredient in beverage base; bad-tasting water will produce a drink with an off-flavor. Spring water is best, if you can get it. If using chlorinated or fluoridated water, let it sit in an uncovered container for several days to "exhaust" most of the chemicals before using. You will need **11 cups water for every 6 cups of beverage base.**

EQUIPMENT

Spoon—large, long-handled
Measuring Spoons—a variety of sizes
Measuring Cups—both wet and dry measures are needed
Colander—large enough to strain the beans
Blender—A blender with a steel or glass bowl is best, as the water will be near boiling when added to the blender. The blender must have a top.
Containers—2-quart total capacity
Cooking Pot—3-quart
Double Boiler—2-quart. One pot inside of another will work.
Cheesecloth—fine-weave cotton cheesecloth (obtainable at fabric stores), two pieces, each 32 inches by 22 inches
Press and Box—Rodale's pot-mount or table-mount press and tofu box

PROCEDURE

1. **Bring 4 cups water to a full rolling boil.**
2. **Add 1/4 teaspoon baking soda** to the boiling water; stir it in quickly.
3. **Add 1 cup dry soybeans** to the boiling water. It is important that the water be boiling when the beans are added. After adding the beans, remove the pan from the heat.
4. **Cover the pot,** and let beans soak at room temperature 8 to 16 hours.
5. **Pour the beans through a colander** and rinse them after they have soaked the required time. The beans should almost triple in volume.
6. **Bring 7 cups water to a full rolling boil.** You may want to add a small additional amount of water to make up for any that may have boiled off. While waiting for the water to boil, continue to next step.
7. **Set up the press and tofu box.** The box must be lined with cheesecloth. First wet the cloth, then line the inside of the box with both pieces, leaving the extra cloth overhanging the top of the box.
8. **Run hot tap water into the blender container to warm it.** It is important that the beans and water be near boiling when blended. Prewarming the blender container will help maintain heat, but more important, it will prevent breakage of the blender bowl.
9. **Blend the soaked beans** and the hot water together. Unless you have a very large blender, you will have to do this in batches. First put 1 cup of soaked beans in the warmed blender container, then add slightly more than 2 cups of hot water. Keep the remaining water on the heat and near boiling. As soon as you add the water, cover the blender and turn it on at low speed, increasing the speed quickly as the mixture blends. Blend at top speed for 1 minute. Don't overblend or your finished beverage will have too much suspended soybean pulp in it. When one batch is blended, put it into a container and repeat process until all the beans and water are used.
10. **Pour the blended mixture into the cheesecloth in the tofu box.** The beverage base will drain through the cloth, and the okara (soybean pulp) will be strained out. Be sure you have a container below the press large enough to hold all of the beverage base. After pouring the slurry into the cheesecloth, stir the pulp in the cloth, fold the cloth over the pulp, and put the pressure plate on top of the cheesecloth. Attach the handle and position the plunger on the pressure plate. Use all of your strength to press the pulp as hard as you can. The pulp will mat together, and you will have to remove the pressure plate, stir and shift the pulp, and re-press it several times to extract all of the beverage base. Work the pulp with the press until you have expressed all of the milk.
11. **Simmer the beverage base for 20 minutes.** A double boiler is best for this, to prevent scalding and burning the milk. However, if a double boiler is unavailable, you can improvise by putting water in the bottom of a large pot and heating the water, then putting the beverage base in a smaller pot and letting it float on the water in the other pot. The goal is to prevent the pot holding the beverage base from coming into direct contact with the heat.
12. The pulp left in the cheesecloth is okara, and may be used as is or dried for future use. See page 33 for details on using okara.
13. **Put beverage base into a container with a cover and refrigerate.**
14. **Clean all utensils well.**

Making Tempeh

INGREDIENTS

The ingredient amounts that are given will make one batch (two 9 x 9-inch pans) of tempeh, weighing a total of $1^1/_2$ to $1^3/_4$ pounds.

Roasted Soy Halves—These are dehulled and roasted soybean halves. Be sure to use unsalted soy halves. The use of roasted soy halves saves the trouble of cooking and dehulling the beans at home, a very time-consuming process. Using commercially roasted soy halves adds slightly to the cost of your tempeh but saves almost 3 hours of preparation time and improves both the flavor and texture. You will need **4 cups soy halves.**

Vinegar—Two types of vinegar can be used: distilled white vinegar or rice vinegar. You will need **1 tablespoon vinegar.**

Starter—The beans must be inoculated with the desired mold. The entire process of making tempeh is dependent upon creating the specific conditions needed for the **Rhizopus oligosporus** mycelium to grow. There are literally thousands of types of mold spores in the air at all times, so care must be taken throughout the process to introduce and encourage only the desired mold. Starter may be purchased or made, as described on pages 20-21. **Use the amount of starter recommended by the manufacturer.**

Water—The purer your water, the better. The water need not be distilled, as it will be boiled. If you must use treated water, let it stand for several days for the chemicals to "breathe" out of it. You will need **8 cups water.**

EQUIPMENT

All equipment must be very clean during the preparation of tempeh. See page 19 for information on cleaning equipment for tempeh production.

Spoon—large, long-handled
Rubber Spatula—any size flexible plastic scraper
Measuring Spoons—a variety of sizes
Measuring Cups—both wet and dry measures are needed
Cooking Pot—4-quart
Colander—large enough to drain 4 cups of soaked beans
Toweling—some form of lint-free toweling or paper toweling
Cookie Sheet—large enough to spread out the soy halves to dry
Mixing Bowl—2 quart
Tempeh Pans—two 9 x 9-inch glass or metal containers. The containers must fit inside the incubator and still permit good air circulation through the incubator. The sides should not be higher than about $1^1/_2$ inches. Metal pans with holes in the bottom, as described on page 19, are best.
Flexible Covering—aluminum foil or plastic wrap
Sharp Object—large needle or sharp nail
Thermometer—A standard outdoor thermometer is best; an oven thermometer won't work.
Incubator—Rodale's tempeh incubator is specially designed to reach and maintain the ideal temperature of 92°F needed for tempeh incubation.

PROCEDURE

All equipment should be very clean before making tempeh.

1. **Boil 8 cups water** in the 4-quart pot.
2. **Add 1 tablespoon vinegar** to the water.
3. **Add 4 cups soy halves** when the water reaches a rolling boil, stirring well to be sure none of these beans stick to the side or bottom of the pot.
4. **Simmer the soy halves** 10 minutes. Wait until the water returns to a boil, then turn down the heat and simmer the soy halves, stirring frequently.
5. **Calibrate the incubator** while the soy halves are simmering. Plug in the incubator, put a thermometer inside, and let it warm up. Wait until the lights go out, then check the temperature. If the temperature is below 92°F, adjust the thermostat until the lights come on; this will add more heat and maintain a higher temperature. If the incubator is too hot, adjust the thermostat the other way to reduce the temperature. For more details on calibrating the incubator, see page 20.
6. **Drain the soy halves** after they have simmered. Pour the beans through a colander to drain off the water. Use a rubber spatula to get all beans out of the pot. Let the beans drain well.
7. **Cool and dry the soy halves** on a cookie sheet that is covered with toweling. Use the spoon and spatula to spread the soy halves evenly on the cookie sheet. Turn the beans once or twice to be sure all beans are fairly dry and have cooled to room temperature; this should take 15 to 20 minutes. Remember, to avoid contamination, don't use your hands.
8. **Inoculate the beans** with the starter. Using the spoon, put the beans in the mixing bowl. Add the recommended amount of starter to the beans, and mix it in well. If the beans are too hot, too wet, or too dry at this stage, the starter will not grow well. Be sure the beans are at room temperature and not too wet or too dry.
9. **Fill the incubator trays** with the inoculated beans. Pack the beans firmly with the spatula; they should form a layer $^1/_2$ to 1 inch thick.
10. **Cover the trays** with aluminum foil or plastic wrap, and poke a series of holes through the cover so the incubating tempeh can breathe. Holes should be punched in a grid pattern $^1/_2$ to $^3/_4$ inch apart.
11. **Clean all equipment.**
12. **Incubate the tempeh** in the incubator for 22 for 24 hours. You should have made sure the incubator was correctly calibrated before putting the trays in for incubation. Not much will happen to the beans for the first 12 hours, but after that, if you peek, you should see a fine, white mold beginning to grow and spread throughout the trays. If you see a lot of moisture condensing on the top and sides of the incubator, wipe it with a clean cloth. Too much moisture in the incubator can spoil the incubation process. Don't open the incubator too often.
13. **Remove the tempeh** from the incubator when the trays are entirely covered with a fine, white mold. Pages 18-19 go into detail on how to tell when tempeh is done.
14. **Store the tempeh** in a covered container in the refrigerator or freezer.
15. **Clean the incubator** and incubation trays.

2 Using Soyfoods

Soybeans offer a veritable smorgasbord of foods. Those we'll discuss here include beverage base, fresh and clabbered soymilk, fresh and frozen tofu, okara, and, last but not least, tempeh. In a kind of geometric progression, each of these foods, in turn, gives a variety of performances—appearing in the form of breakfast, lunch, or dinner foods; as snacks; or even as desserts.

In other words, once you learn to process soy beverages, tofu, and tempeh, you have the makings of an entire cuisine. Because these foods are relatively new to the United States, there are no old family recipes from which to draw. Grandma's okara cookies are unheard of, unless of course, your grandma is from Japan. How to deal with these new foods can be a real mystery. Should you freeze, can, or dry them? Can a baby eat them? What do you do with leftovers? How long will they keep in the refrigerator?

The thing you are most likely to have the most trouble getting used to is the feel of these foods. They have a texture that is vastly different from traditional American foods, especially in their uncooked state. Tempeh, with its soft, velvety feel, is the food most alien to the touch. You'll have to work to get familiar with it.

For the past year the Rodale Test Kitchen has worked extensively with soyfoods, not only developing the recipes on the following pages, but gaining insights into ways to use these fine foods. Day after day, they pressed, slurried, incubated, cooked, stored, dried, froze, baked, mashed, and otherwise manipulated the noble bean into one form or another. They learned from their failures as well as from their successes, so we now are able to provide you with the nitty-gritty information about living with soyfoods that usually would come from Mom and/or Betty Crocker.

Our Test Kitchen Supervisor, Linda Gilbert, sums the message up best, saying "it's important to prepare these foods often, so you gain confidence in your craft. Ease of preparation comes with familiarity."

All the recipes we offer you have been developed especially for this book. They are a good selection of beginning recipes. Once you have developed a taste and a feel for using soyfoods, consult one of the many soy cookbooks available. We feel one of the best for Americanized taste buds is *The Tofu Cookbook*, Emmaus, Pa.: Rodale Press, 1979.

But now, let's start learning how to use these wondrous foods.

SOY BEVERAGES

There are two types of soy beverages, soymilk and the product we've developed, beverage base. The basic difference between soymilk and beverage base is flavor. Soymilk has a soy flavor—beany, somewhat strong, not pleasant—that is usually masked with flavorings and sweeteners. Beverage base is a blander drink, with almost no beany notes. Generally, people are more apt to enjoy beverage base than soymilk. You'll find soymilk best suited for making tofu.

Soymilk and beverage base have a longer shelf life than regular milk and will keep for 12 days or more in the refrigerator in a covered container. And, like milk, they will curdle if boiled and will stick to the bottom of the pan if overheated. Unlike dairy milk, beverage base can be concentrated by using less water during preparation.

Everyone knows what sour milk smells and tastes like. However, if soymilk is something new in your life, you may wonder if you will know when it has soured (clabbered). In storage, soymilk can separate over time, with the heavier element sinking to the bottom of the storage jar. Beverage base will settle even more than soymilk. This settling is natural and is not a sign of souring. Simply shake the jar well before serving. However, if the milk has turned sour, it will form a light yellow curd in a clear yellow whey. If your soymilk has advanced to this stage, it isn't necessary to throw it away. Simply use it as clabbered milk (described later). Beverage base will not sour; it will spoil. You'll know when it is bad.

If you have prepared a large quantity of soymilk or beverage base, and feel you may not use all of it within its 12-day storage period, you can freeze a portion for later use.

Our first experiments with freezing beverage base were not encouraging. However, we learned that by adding two tablespoons of honey to one quart of beverage base, we could freeze it successfully. When the milk thaws, you may notice that a few fine strands of solids have formed. These will disappear when the milk is blended and pose no problem. Freeze soymilk the same way as beverage base, but don't try to make tofu with the thawed milk.

We also found it was not always necessary to thaw the beverage completely to use it. Only partially thawed, it makes a great base for slushes. These icy drinks are fabulous summer refreshers. Just blend the semi-thawed soymilk or beverage base with fresh or frozen fruit. If you're out of fruit, substitute a dash of vanilla. These drinks are thick and satisfying. Nobody would ever guess they are also low in calories.

We should emphasize again—soymilk has a distinctive flavor. This greatly affects the way you use it in place of regular milk in recipes. If you like it, or have become accustomed to it, you will have no problem replacing dairy milk with soymilk in your favorite recipes. Simply substitute on a cup-for-cup basis.

Beverage base has no objectionable beany flavor to be masked. It is much easier for traditional taste buds to accept beverage base directly in place of regular milk with no added flavorings.

In addition to a change in flavor, soymilk and, in this case, beverage base also have a "mouth feel" different from that of dairy milk. It is thinner and somewhat flat compared to cow's milk. Because soy beverages lack the creaminess of dairy milk, you may want to adjust the ingredients of a recipe when you plan to substitute. It is possible to add a few drops of vegetable oil or to combine dairy milk with soy beverages to achieve a creamier texture. You can also combine beverage base with shredded coconut for a sweet, creamy milk. It has a bright white color and a bit higher fat content. An added plus is that the drink will not lose its whiteness during storage.

Soymilk and beverage base both can be successfully used in creamed soup recipes, as long as the soups are well seasoned and filled with vegetables, meat, or fish. If you use soymilk in sauces and soups, try adding a small amount of an acidic ingredient like tomato paste or puree, lemon juice, or vinegar. These sharp flavors really spark up the taste. When you add these acidic ingredients to the soymilk, drizzle in only very small quantities at a time, to prevent curdling.

The recipes that follow are basic and can be adapted to your own tastes with a simple change of flavorings or seasonings. The beverages should be prepared fresh daily. If stored overnight, their color will darken.

Sweetened Soydrink

4 cups beverage base
2 teaspoons vanilla extract
1/4 cup honey or maple syrup

Combine the beverage base, vanilla, and honey or maple syrup, and mix well. Serve chilled.

Yield: 4 1-cup servings

Carob-Banana Shake

2 cups beverage base
4 very ripe medium-size bananas
2 tablespoons honey
2 teaspoons carob powder
1 1/2 teaspoons vanilla extract

Combine the beverage base, bananas, honey, carob powder, and vanilla in a blender, and puree well. Serve chilled.

Yield: 4 1-cup servings

Coconut Soydrink

3 cups beverage base
1 1/2 cups unsweetened dried coconut

1. Combine the beverage base and coconut in a small saucepan. Bring to a boil over medium-high heat, stirring frequently. Remove from the heat and cool.
2. Puree the mixture in a blender. Strain out the solids through a fine-meshed strainer. Chill.
3. Enjoy this drink plain, or use it as a base for other drinks. Flavor with fresh or dried fruit, carob, or other flavorings. This drink will not darken with storage; it will stay very white for a long time.

Yield: 2 cups

Strawberry-Banana Drink

2 cups beverage base
2 cups frozen strawberries
1 very ripe medium-size banana
5 tablespoons honey
1 teaspoon vanilla extract

Combine the beverage base, strawberries, banana, honey, and vanilla in a blender, and puree well. Serve chilled.

Yield: 4 1-cup servings

Piña-Colada Shake

2 cups beverage base
1 cup dried pineapple chunks
2 cups Coconut Soydrink (see recipe above)
1/2 cup pineapple juice

1. Combine the beverage base and dried pineapple in a small saucepan. Bring to a boil over medium-high heat, stirring frequently. Remove from the heat and chill.
2. Puree the pineapple and beverage base in a blender. Add the coconut soydrink and blend well. Thin to desired consistency with pineapple juice.

Yield: 5 1-cup servings

Banana-Maple Iced Dessert

3 cups beverage base
3/4 cup maple syrup
2 teaspoons vanilla extract
1/2 cup heavy cream
1 1/2 cups very ripe banana slices

1. Combine all ingredients in a blender, and blend well.
2. Process in ice cream maker.

Yield: 1 quart

Cream Soup Base

4 tablespoons butter
4 tablespoons whole wheat flour
2 cups beverage base

1. Melt the butter over low heat in a 2-quart saucepan. Stir in the flour and mix until smooth. Cook gently, stirring constantly, for 2 to 3 minutes.
2. Slowly pour in the beverage base, stirring with a wire whisk to prevent lumps. Bring to a simmer over low-medium heat and simmer for 2 more minutes. This base may be used for any cream soup.

Yield: 2 cups

Cream of Chicken Soup

2 tablespoons butter
1 celery stalk, thinly sliced
2 carrots, thinly sliced
1 large onion, coarsely chopped
1/4 teaspoon dried thyme
2 teaspoons chopped fresh parsley
1 cup chicken stock
2 cups Cream Soup Base (see recipe above)
1/2 cup chopped cooked chicken meat
black pepper
freshly grated nutmeg

1. Melt the butter in a 2-quart saucepan. Sauté the celery, carrots, and onion over medium heat until tender. Stir in the thyme and parsley.
2. Add the chicken stock and simmer over medium heat, covered, stirring frequently, 15 to 20 minutes. Meanwhile, prepare the cream soup base.
3. Slowly stir the stock into the warm soup base. Do not let the soup come to a boil. Season to taste with pepper and nutmeg. Serve hot.

Yield: 4 1-cup servings

CLABBERED SOYMILK

When soymilk turns sour, which will happen after 12 to 14 days of refrigerator storage, it can be used in place of buttermilk in a wide variety of recipes. Beverage base will not clabber; it will go bad, but it won't sour.

Clabbered soymilk has a tantalizingly sour flavor and a soft, fragile consistency. It will no longer look like milk but will be a soft, yellow curd in a clear, yellow whey.

To use clabbered soymilk as a cup-for-cup substitute for buttermilk, combine the curd and whey by stirring.

To use clabbered soymilk for pudding, custard, cream cheese, sour cream, and whipped desserts and toppings, first drain the whey from the curds. If you use a jelly bag, no squeezing or pressing is necessary.

Whipped Blueberry Dessert

1 tablespoon gelatin
1 1/4 cups drained clabbered soymilk
2 tablespoons honey
1 teaspoon vanilla
1 cup blueberries

1. Dissolve the gelatin in 1/4 cup of the clabbered soymilk over very low heat.
2. Add the gelatin mixture to the remaining clabbered soymilk in a large glass bowl. Mix well.
3. Refrigerate until the mixture begins to thicken, about 30 minutes. Then beat until foamy. Add the honey, vanilla, and then the blueberries. Whip until mixture is smooth and foamy and the fruit is incorporated into the mixture.
4. Serve cold.

Yield: 4 small servings

Soymilk Cream Cheese

2 cups drained clabbered soymilk
1/2 cup heavy cream
4 cups milk

1. Combine all ingredients in a large stainless steel or enameled pot or bowl.
2. Slowly heat to between 100° and 102°F. Let stand in a warm place (85° to 100°F), uncovered, for about 24 hours or until curds form.
3. Line a colander with muslin or several layers of cheesecloth; pour the mixture through gently.
4. Bring the ends of the cloth together to form a bag, tie it, and suspend bag from the faucet spout to drip into the sink. Hang for about 16 hours or until the cheese is set.

Yield: 1 1/2 cups cheese

Sour Cream

2 cups drained clabbered soymilk
1 cup farmer's cheese

Blend soymilk and cheese together in a blender until smooth and creamy.

Yield: 3 cups

TOFU

As you know, tofu is a soft cheese made from soymilk. Its consistency is smooth and fragile. Like other soy products, it is low in calories and fat but high in protein.

The most fascinating aspect of tofu is its versatility of form. By manipulating the water content as you prepare it, you can customize its consistency. You can create a product that is smooth and as creamy as yogurt or firm enough to be sliced and stir-fried. Each varying degree of firmness is best suited for its own type of recipe—the smooth, creamy tofu for dips, the firm tofu for breading and frying.

Unlike any other food, tofu must be stored in water. Keep tofu in a closed container covered with fresh water, and store it in the refrigerator. If you change the storage water and rinse the tofu daily, it will last for at least ten days.

Despite being stored in water, as each day in storage passes the tofu loses a little water. It becomes firmer in texture and stronger in flavor.

FRESHENING STORED TOFU

If the tofu begins to sour, it can be refreshened. This process will cause the tofu to become denser, firmer, and drier. It will also lose some flavor. However, in its firmer form, it will be even more suitable for some recipes than it was when freshly made.

To freshen tofu that has started to sour or to warm tofu for serving with sauce, bring six cups of water to a boil in a 3-quart saucepan. Add 1 pound of tofu, reduce the heat to low, and cover the pot. Allow the tofu to simmer for about 2 minutes. Remove it from the water with a slotted spoon. The longer the tofu cooks, the firmer it will become. Also, the smaller the pieces of tofu, the firmer they will become.

FREEZING TOFU

For long-term storage, it is best to freeze tofu. Freezing changes the quality and texture of tofu after thawing, making it more suitable for some dishes than for others. Its texture after thawing is spongy and chewy. It will absorb oil, making it difficult to sauté or stir-fry. However, that absorbent quality makes frozen tofu very desirable in stews and in dishes where small pieces can absorb the spicy sauce and offer a meatlike chewing texture.

To freeze tofu, first rinse it with fresh water and pat dry with paper towels. You can freeze tofu in any form you want: slices, cubes, or in the original cake. Place the tofu in freezer bags, squeeze out the air, and freeze. Frozen tofu is good for about six months. After 48 hours, the desired change in texture will have occurred.

For the quickest thaw in the West, steam the tofu. However, if you don't want precooked tofu in your recipe, simply thaw it in a colander at room temperature. Freshly thawed tofu makes a very good meat extender when the moisture is squeezed out and it is crumbled, seasoned, and bound together with eggs. It has a texture very similar to ground beef.

It is not always necessary to thaw frozen tofu before using it. For instance, if you plan to use it as an ingredient in stew, don't defrost it; simply add 15 minutes to the cooking time.

The key to success when cooking with tofu is to understand and exploit its great adaptability and versatility of form. Soft tofu has a high water content and is creamy. If the dish you are preparing calls for tofu with a custardlike texture, then soft tofu is what you want. It blends and purees well. And when it is pureed, it becomes very smooth and is easily used for making cheese spreads, dips, salad dressings, puddings, and custards. It can be substituted for ricotta cheese in recipes or replace sour cream or yogurt in a cake and pastry filling.

Firm tofu has a low water content and holds its shape well. It is ideal for slicing and is excellent used in casseroles, stir-fry dishes, soups, salads, and sautés; served with sauce; stuffed, marinated, breaded, fried, or made into tempura; or as the dipping ingredient in fondues or dips. Firm tofu can be beaten with an electric mixer. It will not become creamy, but rather will break into uniform, dry crumbs. In this form, tofu can be combined with meat or cheese to extend its quantity. In baked goods, stuffings, and fillings, the crumbs will take the place of eggs or milk. In these dishes the tofu crumbs will not act as a binder, the way eggs do, but will offer substitute protein and contribute to the texture of the finished dish.

Some recipes specify whether soft or firm tofu should be used. If the recipe calls for firm tofu when you have only the soft kind, simply re-press the soft tofu to remove some of its moisture before proceeding. If you need soft tofu but have only firm, you can use it in the recipe, but with less success. You may have to add a bit of additional liquid to the recipe in order for firm tofu to be substituted for soft tofu successfully.

Many recipes will require a specific treatment of tofu, specifying that it is to be mashed or beaten or in some other form. To prepare the tofu in different ways, refer to chart 2-1 for guidelines. As you prepare tofu in different ways, its value to a recipe changes. Chart 2-2 gives the equivalents for tofu's different forms. Use it when substituting forms in a recipe.

In addition to the many recipes that follow, here are some general guidelines to follow when cooking with tofu.

TOFU AND CHEESE

To extend a cheese, mix equal parts of grated cheese with beaten firm tofu. Swiss, Parmesan, cheddar, colby, cream, or cottage cheese all work well; experiment with others. Refrigerate overnight to allow the flavors to

Forms of Tofu

Type of Tofu	Procedure
Mashed	Break or crush the cake to soften the tofu. Don't be afraid to use your hands. Mashing makes the tofu easier to puree in a blender.
Beaten	First mash the tofu, then beat it with an electric mixer (not a blender). Beat the tofu until it has achieved a dry, crumblike consistency.
Squeezed	By removing the moisture, tofu becomes firmer and more cohesive. Cut the tofu into small chunks, wrap them in a clean, cotton dish towel, and twist to remove as much water as possible.
Pressed	For firm tofu that remains in cake form, return the tofu to the press. Hang a 3- to 5-pound weight on the cup hook and press for 1/2 hour.

Chart 2-1—Tofu has many forms; each is best suited to a specific type of dish. One form of tofu can usually be transformed into another form.

blend. Add a small amount of grated Parmesan cheese when mixing tofu with another type of cheese, to accent the original cheese's flavor. Use the mixture as you would the original cheese.

For cheese fillings, combine cottage or ricotta cheese with pureed soft tofu in equal proportions. Use in pastries, cakes, casseroles, lasagna, quiches, or omelettes. Again, add a small amount of grated Parmesan to bring out more cheese flavor.

TOFU IN MEAT DISHES

Firm tofu made with lots of spices makes an excellent sausage substitute in casseroles.

Tofu Equivalent Measurements

(1 lb. firm tofu)

Form	Amount
1″ Cubes	2 2/3 cups
Mashed	2 cups
Cake	7″ × 4″ × 1″
Squeezed	1 1/4 cups
Beaten	1 3/4 cups

Chart 2-2—When transforming tofu from one form to another, check this chart for equivalent measurements.

Mashed or beaten firm tofu makes a good extender for ground beef. Use about 2 cups of tofu to ½ cup of ground beef. If you plan to broil these burgers, do not use lean ground meat. There will not be enough fat for proper cooking.

Substitute tofu cubes for half or all the meat chunks in almost any recipe. Marinate the tofu chunks in a brown marinade. If the recipe requires long simmering, add the tofu cubes during the last 10 minutes of cooking to retain its tenderness. Tofu will better hold its shape in stews and casseroles if it is first cubed and sautéed.

Extend dips and spreads by blending in soft tofu. Increase the seasonings proportionately. When making mayonnaise-based dips or spreads, you can reduce the required amounts of eggs and oil by half, and replace them with an equal quantity of mashed tofu.

Thicken sauces with tofu. It acts by adding solids to the sauce and increases the protein value. It will not act as a starch or protein thickener; the reaction is more like adding pureed vegetables to a sauce.

When making a soufflé, substitute beaten tofu for half the eggs. Add the beaten tofu to the custard base, then fold in the egg whites. Extra egg whites will make the soufflé lighter.

Make puddings with tofu. Just puree soft tofu with fruit, carob, vanilla, or any other favorite flavoring. Chill and serve the combination.

With these guidelines, you can experiment to your heart's content. If you prefer to follow a recipe, we offer these below.

Frozen-Tofu Meatballs

1 onion, finely chopped
2 garlic cloves, minced
1/4 cup olive oil
1 teaspoon dried oregano
1 cup frozen tofu, thawed, pressed, and crumbled
1/2 teaspoon soy sauce
2 teaspoons tomato paste
1 egg, lightly beaten

1. Lightly sauté the onion and garlic in 2 tablespoons of the olive oil.
2. Remove from heat and stir in the remaining ingredients. Mix to combine well.
3. Form the mixture into 16 small balls.
4. Sauté until browned in the remaining olive oil.
5. Simmer for 15 minutes in your favorite tomato sauce.

Yield: 4 servings

Tofu and Pineapple Pastry Filling

1 1/2 cups mashed tofu
1 1/2 cups coarsely chopped dried pineapple
1/2 cup almonds, ground
3 tablespoons soymilk
1/4 teaspoon cinnamon

Combine all ingredients in a blender, and puree. Chill.

Yield: 2 cups

Quick Blender Tofu Soufflé

1 cup milk
1 1/2 cups mashed tofu
1 carrot, coarsely chopped
1 scallion, coarsely chopped
1/2 cup grated Parmesan cheese
1 teaspoon dried dillweed
1/4 teaspoon dry mustard
2 tablespoons butter, melted
4 egg yolks
3 slices whole wheat bread, broken into small pieces
4 egg whites, beaten until stiff

1. Preheat oven to 325°F.
2. Combine the milk, tofu, carrot, scallion, cheese, dillweed, mustard, butter, egg yolks, and bread pieces in a blender. Puree until smooth. Fold in the egg whites.
3. Pour into an oiled soufflé dish.
4. Bake in preheated oven for about 50 minutes.
5. Serve with a tomato or herb sauce. Melted cheese is also very good.

Yield: 4 to 6 servings

Variation

For a more custardlike soufflé, use 1 less egg and 1/2 cup less milk. This soufflé will have a firmer texture and will store well for reheating.

Scalloped Tofu au Gratin

4 tablespoons butter
1 large onion, thinly sliced
1 small pepper, thinly sliced
3 tablespoons whole wheat flour
1 cup milk
1/4 teaspoon white pepper
1 teaspoon dry mustard
1 1/4 cups grated cheddar cheese
3 1/2 cups 1/4-inch squares tofu
2 tablespoons soft whole grain bread crumbs

1. Melt 1 tablespoon of the butter in a skillet. Sauté the onion and pepper until tender. Reserve.
2. In the same skillet, melt another tablespoon of the butter, and lightly sauté the tofu for about 5 minutes.
3. Melt the remaining butter in a small saucepan, and stir in the flour. Cook lightly for several minutes, stirring constantly until smooth and bubbling. Slowly stir in the milk and cook on low heat, stirring frequently, until thick and beginning to bubble.
4. Remove the sauce from the heat, and stir in the white pepper, mustard, and 3/4 cup of the cheese.
5. Butter a shallow 9-inch oven proof casserole dish. Layer first with half of the tofu, cover with half of the sauce, then half of the vegetables, then the remaining tofu, sauce, and vegetables.
6. Top with bread crumbs and remaining cheese.
7. Bake in a preheated 350°F oven 20 minutes or until bubbly and browned on top.

Yield: 4 servings

Garlic Herb Cheese

4 garlic cloves, coarsely chopped
2 tablespoons coarsely chopped mixed fresh herbs (for example, basil, chervil, chives, dill, lemon thyme, scallions, sorrel, or tarragon)
1 1/4 cup coarsely chopped tofu
4 ounces cream cheese, softened
1/4 cup butter, softened
2 tablespoons sour cream or yogurt

Garnish

freshly ground black pepper or additional chopped herbs

1. On a cutting board, combine the garlic, parsley, and herbs, and mince.
2. Wrap the tofu in a clean cotton towel, and wring out the moisture.
3. Combine the seasonings mixture, tofu, cream cheese, butter, and sour cream in a large bowl. Beat until smooth with an electric mixer.
4. Chill until firm. The cheese may now be shaped as desired and rolled in the pepper or herbs to garnish.
5. Refrigerate overnight before serving, to blend the flavors.

Yield: 1 cup

Pesto Tofu Cheese

1 large garlic clove
1/2 cup fresh basil leaves, loosely packed
1 tablespoon finely chopped fresh parsley
1/4 cup sunflower seeds or pignolias (pine nuts)
1 tablespoon lemon juice
2 tablespoons sour cream or yogurt
1 cup coarsely chopped tofu
1/4 cup grated Parmesan cheese

Garnish

2 tablespoons finely chopped sunflower seeds or pignolias

1. On a cutting board, combine the garlic, basil, parsley, and sunflower seeds and mince.
2. Combine the lemon juice and sour cream in a blender. Add the basil mixture and blend briefly (4 to 5 seconds).
3. Wrap the tofu in a clean cotton towel, and wring out the moisture.
4. Add the tofu and cheese to the mixture in the blender. Blend until smooth.
5. Chill until firm. The cheese may now be shaped as desired and rolled in the chopped sunflower seeds to garnish.
6. Refrigerate overnight before serving, to blend the flavors.

Yield: 1 cup

Dill Tofu Spread or Dip

4 egg yolks
1/2 cup olive oil
2 garlic cloves, coarsely chopped
1 tablespoon minced fresh dillweed or 1 teaspoon dried dillweed
2 tablespoons finely chopped fresh parsley
1/2 teaspoon paprika (Hungarian, if available)
1/2 teaspoon dry mustard
1/2 teaspoon soy sauce
2 tablespoons lemon juice
3/4 cup mashed soft tofu

1. In a bowl, beat the egg yolks with a wire whisk until they begin to thicken. Slowly, by drops as for mayonnaise, whisk in 1/4 cup of the olive oil. Beat constantly with the wire whisk. The mixture should be smooth and have the consistency of mayonnaise. Transfer to a blender.
2. Add the garlic, dillweed, parsley, paprika, and mustard. With the blender on low, gradually add the remaining oil in a thin stream. When the oil is completely incorporated into the mixture, slowly add the soy sauce, lemon juice, and tofu. Blend until smooth and refrigerate.
3. If your mayonnaise should thin out or separate, warm a glass or stainless steel bowl and a wire whisk in hot water. Dry thoroughly. Lightly beat 1 egg yolk in the bowl. Gradually, drop by drop, beat in the broken mayonnaise.
4. It is best to refrigerate this spread overnight to blend the flavors.

Yield: 1 1/3 cups

Broiled Tofu Burgers

2 cups mashed tofu
1/2 pound ground beef (not lean)
2 tablespoons soy sauce
2 teaspoons cider vinegar
2 garlic cloves, minced
1/4 teaspoon black pepper
1/4 cup tomato paste

1. Mix all ingredients together well, using a large spoon or your hands. Set aside for about 30 minutes.
2. Form into 12 3-inch patties, and broil for 5 to 7 minutes on each side, or until browned.
3. Serve as you would a hamburger—as an entree, with sauce, or on a bun. Note: The fat in the beef is necessary for broiling these burgers. If you use lean ground beef there may not be enough fat for broiling.

Yield: 4 servings

Carob Tofu Pudding

2 cups 1-inch cubes soft tofu
1/2 cup cold soymilk (dairy milk may be substituted)
1 teaspoon vanilla extract
3 medium-size bananas, peeled, frozen, and cut into 2-inch pieces
4 tablespoons carob powder

Puree all ingredients in a blender until smooth. Serve well chilled.

Yield: 4 to 6 servings

OKARA

Of all soyfoods, okara is the most magical. Okara offers fiber and protein and can be included in a wide variety of foods without making its presence known. It is a by-product of making tofu, and, although it has lots of uses, it is often overlooked.

You can use okara fresh when it has a consistency similar to fresh, grated coconut, or you can use dried okara powder. For powder, dry the okara in a food dehydrator, then grind it in a blender or food processor. It stores well and has a long shelf life. It freezes well; just bag it and freeze it.

If you're not acquainted with the many uses of okara, you're in for a treat. The following recipes for okara will show you what is possible with this magical by-product of tofu.

Okara Pudding with Raisins and Cinnamon

1/2 cup finely ground dried okara
2 cups milk
1 egg, beaten
2 tablespoons honey
1/2 teaspoon vanilla extract
1/2 teaspoon cinnamon
1/2 cup raisins

1. Combine okara and milk in a 1-quart saucepan. Simmer over medium heat, stirring frequently, 20 minutes.
2. Beat egg with honey. Add some of the hot mixture to the egg and honey, then slowly pour egg mixture into hot milk, stirring constantly.
3. Cook over low heat until thickened and almost ready to boil. Remove from heat. Add vanilla, cinnamon, and raisins. Serve hot or cold.
4. This is a tasty, high-fiber breakfast cereal. The okara acts as a thickening agent and provides protein and iron.

Yield: 3 cups (4 to 6 servings)

Okara Cheese Crackers

4 tablespoons butter
about 5 tablespoons whole wheat flour
1/8 teaspoon cayenne powder
1/2 teaspoon paprika
1/4 teaspoon caraway seeds
2 tablespoons grated Parmesan cheese
1/3 cup grated cheddar or Swiss cheese
1/2 cup okara
2 teaspoons raw sesame seeds

1. Cut the butter into the flour in a medium-size bowl.
2. Add the cayenne powder, paprika, caraway seeds, Parmesan cheese, and cheddar cheese. Knead lightly to combine ingredients.
3. Add the okara and knead until well combined. The mixture should be uniformly colored. If it is too sticky to work, add a bit more flour.
4. Chill, covered, at least 1 hour.
5. Roll the dough out on a floured board to approximately 1/8 inch thick. Sprinkle with the sesame seeds. Lightly press the seeds into the dough.
6. Cut the dough into desired shapes and place them on a lightly oiled cookie sheet.
7. Bake in a preheated 400°F oven until lightly browned, 12 to 15 minutes.
8. Cool the crackers on a wire rack.

Yield: 3 to 4 dozen crackers

Okara Patties (basic recipe with variations)

1 small onion, finely chopped
1 celery stalk, finely chopped
1 large garlic clove, minced
2 tablespoons butter
1 medium-size carrot, grated (about 1/2 cup)
2 tablespoons finely chopped fresh parsley
1 cup soft whole grain bread crumbs
1 cup okara
1 tablespoon tomato paste (optional)
juice of 1 medium-size lemon (about 2 tablespoons)
2 eggs, lightly beaten
1/8 teaspoon white pepper
1/2 teaspoon ground mustard
vegetable oil for frying

1. Sauté onion, celery, and garlic in butter until tender. Transfer to a large bowl.
2. Stir in the carrot, parsley, bread crumbs, okara, tomato paste (if used), lemon juice, eggs, pepper, and mustard. Add seasoning variation, if desired. (See below.)
3. Mix well to combine ingredients. Form into 24 2-inch patties. Heat oil in a skillet. Fry patties over medium heat until browned on each side.

Yield: 4 to 6 servings

Variations

Italian

1/4 cup grated Parmesan cheese
1 teaspoon dried basil
1 teaspoon dried summer savory
1/2 teaspoon dried thyme
1/4 teaspoon garlic powder (optional)

Peanut

4 to 6 teaspoons soy sauce
1 cup raw peanuts, ground
1/4 to 1/2 teaspoon ground ginger

Cheddar Cheese

1/4 cup wheat germ
1/2 teaspoon cayenne pepper
1 teaspoon garlic powder
1/4 teaspoon ground cumin
1/2 cup grated cheddar cheese

TEMPEH

Tempeh is a protein- and fiber-rich food that combines a beneficial mold with legumes. The resultant velvety white cakes can be prepared in a wide variety of ways. Flavors can be changed with the substitution of one legume for another, so that you can dine on hearty soy, chewy peanut, delicate soy-sesame, or fine-textured okara tempeh.

Preparing tempeh is one of life's little pleasures. You'll be amazed at how the white mold turns a crispy golden-brown in the skillet when fried. Cut the tempeh cake into cubes, fry it, then spear each with a toothpick, and serve as a light snack.

Most first-time tempeh cooks make the same simple mistake. They cut the cubes too large, so that when someone takes a bite, a long strand of silvery white mold extends from the cube. Little cubes will be gobbled up quickly, but the larger servings will sit until stone cold. Lesson one: Serve tempeh cubes in bite-size pieces.

Tempeh is best fresh from the incubator. However, once incubated, the mold-covered bean cake can be held in the refrigerator for up to seven days before cooking. When you store uncooked tempeh, be sure it is carefully covered in a clean, sealed container.

For long-term storage, it is possible to freeze tempeh after it has been incubated. The tempeh will lose some flavor, and in addition, the texture becomes somewhat sticky and gummy. Because of this stickiness, it is harder to fry or sauté the tempeh successfully, as it absorbs the oil and sticks to the pan.

You'll find tempeh to be versatile in the kitchen. Your uses for it will grow with your experience. Start with the recipes that follow, and then begin substituting tempeh in your favorite recipes. The critical point is to work with small pieces, so it gets completely cooked and can absorb more of the flavors of the spices and sauces. When using tempeh in soups or stews, dishes with a large proportion of liquid, precook the tempeh for better flavor.

We feature a recipe for dry-fried tempeh. This recipe has a myriad of uses, because it allows you to use leftover tempeh like ground nuts—sprinkled over vegetables, as a topping for casseroles, or even added to a bowl of granola.

Tempeh Stuffed Peppers

4 medium-size green peppers
3 tablespoons vegetable oil
1 garlic clove, minced
3/4 cup finely chopped onion
3/4 cup finely chopped celery
1 cup coarsely chopped mushrooms
1 cup finely chopped tomatoes
1 teaspoon dried chervil
1 teaspoon dried basil
1 1/2 cups finely chopped tempeh
3/4 cup grated colby longhorn cheese
black pepper
1 cup pureed tomatoes

1. Slice the tops off the peppers and remove seeds. Steam gently until barely tender, 5 to 7 minutes.
2. Core and finely chop the green pepper tops. Set aside.
3. Heat the oil over medium heat in a large skillet. Sauté the garlic, onion, celery, and mushrooms until wilted.
4. Add the chopped tomatoes and cook over medium heat, stirring frequently until the liquid dries off. Stir in the chervil and basil.
5. Stir in the tempeh and chopped green pepper. Continue to sauté 10 minutes more. More oil may be needed at this time.
6. Remove from heat, and stir in 1/2 cup of the grated cheese. Season with black pepper.
7. Preheat oven to 350°F.
8. Stuff peppers with mixture, and place them in an oiled, shallow baking pan or dish. Pour pureed tomatoes around, not over, the peppers. Top each pepper with the remaining cheese.
9. Bake in preheated oven for 30 minutes or until peppers are tender.

Yield: 4 servings

Tempeh Piecrust

1 cup whole wheat bread flour
1/2 cup finely ground Dry-Fried Tempeh (see recipe below)
6 tablespoons butter, softened
3 to 4 tablespoons ice water

1. Combine flour and tempeh and mix well. Cut in butter.
2. Sprinkle with water while mixing with a fork. Add only enough water to make the crumbs cling together and form a ball.
3. Knead together for a few seconds. Press into a buttered 9-inch pie pan.
4. The piecrust is now ready for the filling of your choice. Bake according to filling directions. If filling does not require baking, bake crust at 425°F for 8 to 10 minutes.

Yield: 1 9-inch piecrust

Tempeh Bread Pudding

3 cups 1-inch whole grain bread cubes
4 cups milk
1 cup finely chopped tempeh
1/2 cup raisins
4 eggs
1/2 cup honey
1 teaspoon vanilla extract
1/2 teaspoon nutmeg

1. Preheat oven to 350°F.
2. Combine the bread with 1 cup of the milk. Gently mix in the tempeh and raisins.
3. Scald the remaining milk and cool. Transfer the bread and tempeh mixture to an oiled 8-inch-square ovenproof casserole dish.
4. Beat the eggs in a 2-quart bowl. Add the honey, vanilla, 1/4 teaspoon of the nutmeg, and the scalded milk, and beat together well. Pour over the mixture in the casserole.
5. Sprinkle with the remaining nutmeg. Bake in a preheated oven for about 40 minutes or until a knife inserted into the center comes out clean.

Yield: 4 servings

Tempeh Burgers

2 cups finely chopped tempeh
1 onion, finely shredded
1 carrot, finely shredded
1 teaspoon minced garlic
3/4 cup toasted wheat germ
1/2 cup cooked long-grain brown rice
1/4 teaspoon dried dill-weed
1/4 teaspoon dried basil
2 eggs, lightly beaten
vegetable oil for frying

1. Combine tempeh, onion, carrot, garlic, wheat germ, rice, dillweed, basil, and eggs in a large bowl. Mix well. Shape into 12 patties.
2. Fry in oil over medium heat until a dark golden brown on each side.

Yield: 4 to 6 servings

Pumpkin-Tempeh Bread

1/2 cup peanut oil
1/3 cup honey
1/3 cup sorghum syrup or mild molasses
2 eggs
1 1/2 cups mashed cooked pumpkin
2 cups whole wheat flour
1 teaspoon baking powder
1 teaspoon baking soda
1 1/2 teaspoons ground cinnamon
1/8 teaspoon ground allspice
1/8 teaspoon ground cloves
3/4 cup Dry-Fried Tempeh (see recipe below)
1/2 cup raisins

1. Beat oil, honey, and sorghum syrup together with a wire whisk. Alternately beat in eggs and pumpkin.
2. In a separate bowl, combine the flour, baking powder, baking soda, cinnamon, allspice, and cloves. Gradually beat into liquid mixture. Stir in tempeh and raisins.
3. Put into a 9 x 5-inch greased and floured loaf pan, and bake at 350°F for 60 minutes. Turn heat to 325°F, cover top with foil, and bake 15 minutes more.

Yield: 1 9 x 5-inch loaf

Tempeh Chili

1 tablespoon vegetable oil
1 large onion, coarsely chopped
1 green pepper, coarsely chopped
1/2 pound ground beef
1 1/2 cups 1/2-inch to 3/4-inch squares tempeh
2 tomatoes, coarsely chopped
1 bay leaf
1/8 to 1/4 teaspoon dried red pepper flakes
1 to 2 teaspoons chili powder
1 teaspoon soy sauce

1. Warm the oil in a heavy skillet over medium heat. Stir in the onion and green pepper, and sauté until tender.
2. Break up the beef into the pan, and brown over medium heat. Drain off the extra fat.
3. Stir in the remaining ingredients. Mix well and cover. Reduce heat to low and simmer 10 minutes, stirring occasionally.

Yield: 4 servings

Dry-Fried Tempeh

3 cups finely chopped tempeh

1. Heat a heavy skillet, preferably cast iron, over medium-low heat. Stir in the tempeh. Fry the tempeh, stirring and tossing constantly with a metal or wooden spatula to prevent sticking or burning, until well toasted.
2. Grind finely in a blender or coffee mill.

Yield: 1 1/2 cups

Tempeh Parmesan

1 1/2 cups tomato sauce
12 2 1/2-inch by 2 1/2-inch pieces of tempeh
2 teaspoons soy sauce
4 teaspoons water
1/4 cup whole wheat flour
1 egg, beaten with 1 tablespoon water
1 cup fine whole grain bread crumbs
1/4 cup olive oil
1/2 pound mozzarella cheese, thinly sliced or grated
1/4 cup grated Parmesan cheese

1. Spread a thin layer of sauce over the bottom of an oiled 10-inch-square ovenproof casserole. Reserve remaining sauce.
2. Combine the soy sauce and water and mix well.
3. Dip each piece of tempeh into the water-soy sauce mixture, then into the flour, pressing the tempeh firmly into the flour to coat evenly. Then dip each piece of tempeh into the egg and then into the bread crumbs, shaking off the excess.
4. Brown the slices in a thin layer of olive oil in a heavy pan, and place them in the casserole. Spread the mozzarella cheese over the top evenly, cover with the remaining tomato sauce, and sprinkle with the Parmesan cheese.
5. Bake at 400°F until cheese melts (about 15 minutes).

Yield: 4 servings

Fruit and Tempeh Stuffing

1/2 cup finely chopped dried pineapple
1/2 cup raisins
1 cup water
1 tablespoon soy sauce
1 2-inch cinnamon stick
1/2 cup unsweetened dried coconut
2 tablespoons butter
2 cups 1/4-inch pieces soy-sesame tempeh
1 large onion, finely chopped
1 cup day-old whole grain bread crumbs
1/2 teaspoon ground cumin
pinch ground cardamom

1. Combine the pineapple, raisins, water, soy sauce, and cinnamon stick in a 2-quart saucepan. Simmer, uncovered, until the liquid is reduced to about 1 tablespoon. Remove from the heat and transfer to a 2-quart mixing bowl.
2. Remove the cinnamon stick and stir in the coconut.
3. Melt the butter in a skillet, and sauté the tempeh and onion until lightly browned.
4. Add the sautéed tempeh to the fruit mixture. Stir in the bread crumbs, cumin, and cardamom, and mix well.
5. This stuffing may be used for chicken, fish, or vegetables. Extra may be frozen.

Yield: 3 1/2 cups

Tempeh Barbecued Beans

Sauce

1/2 cup tomato paste
1/2 cup water
2 teaspoons soy sauce
1/2 cup finely chopped tomatoes
2 tablespoons cider vinegar
1/2 teaspoon cinnamon
1/2 teaspoon dry mustard
1/4 cup finely chopped onion
1/4 cup finely chopped green pepper
2 tablespoons blackstrap molasses
2 tablespoons honey
cayenne powder to taste
2 cups 1/2-inch to 3/4-inch squares tempeh
1 teaspoon butter

1. Preheat oven to 375°F.
2. Combine the sauce ingredients and mix well.
3. Add the tempeh and stir gently to coat the pieces with the sauce.
4. Pour the tempeh mixture into an oiled ovenproof casserole dish, and dot with the butter.
5. Bake, uncovered, in a preheated oven 35 to 45 minutes.

Yield: 4 servings

Peanut Tempeh Sauce

3 tablespoons vegetable oil
1 small onion, coarsely chopped
1 garlic clove, coarsely chopped
3/4 cup finely chopped peanut tempeh
1 tablespoon unsweetened dried coconut
1 tablespoon honey
2 teaspoons lemon juice
1/2 cup milk (dairy or soymilk)
1/4 teaspoon ground coriander seed
grated rind of 1/2 lemon
soy sauce to taste
cayenne pepper to taste

1. Heat the oil in a medium-size skillet. Sauté the onion, garlic, and tempeh over medium-low heat until the tempeh begins to brown.
2. Transfer tempeh mixture to a blender. Add the coconut, honey, lemon juice, milk, coriander seed, and lemon rind. Puree until smooth.
3. Season to taste with soy sauce and cayenne pepper. Serve hot or cold.
4. This is an appropriate sauce for plainly prepared fish, cabbage, cauliflower, or other cooked vegetables.

Yield: 1 cup

Section II Introduction

The information you've just read should have whetted your appetite considerably, and we hope it has given you a new understanding of the vast potential of soyfoods. This part of the book will give you all the information you need to build equipment to make soyfoods a convenient part of your life. We show you exactly how to build your own tofu box, your choice of two types of presses, and a tempeh incubator.

There are four separate projects in this part of the book; each one is built separately. The first is a tofu box for making tofu. When you are finished with the tofu box you'll be ready to put together one of the two presses designed to go with it. Tofu can be made with just the tofu box, but the addition of a press will make your tofu making much easier. We feel that a press to be used in coordination with the box is almost a necessity. After reading through chapter 1, choose which press you will build: a press to use on top of a canning kettle or a table-mounted press. The two chapters on the presses follow the tofu box chapter. The last project is a tempeh incubator, designed solely for making tempeh.

Each chapter starts out with a cutting phase to show you just how to get all of the pieces needed to assemble the project from the minimum amount of material. Follow the cutting diagrams, and pay close attention to the direction of the wood grain as you cut each piece. After you have cut all of the pieces, you'll be ready to put them together. Generally, it is not a good idea to deviate from the path we've charted for you when it comes to building technique. If we tell you to assemble two pieces a certain way, rest assured that we have tried many other ways and found that particular procedure the best and easiest. For instance, if we tell you to cut a bevel on one piece using a circular saw or saber saw and then suggest planing the bevel on another piece, the change in procedure was well thought out. Using a plane worked better and easier in one case, while cutting with a saw suited the other. If you change the assembly sequence or try to substitute materials, you may run into trouble.

Each piece has a label and a name, both of which can be found on the blueprint sheet detailing that particular project. The label of the piece is comprised of a number and a letter. The number refers to the number of the blueprint sheet where you will find that piece, while the letter identifies the piece more specifically. The blueprint sheets should be studied carefully.

Since each chapter stands on its own, the tools and materials needed for each project are listed at the beginning of each chapter. We have not used stationary power tools, because they are uncommon in the typical home shop, but they certainly should be used if you have them. Otherwise, use the tools we recommend and you will have no trouble.

The materials lists are complete with the exception of finishes to be put on the completed project. The tofu box and presses do not require any finish at all; time has shown that bare wood is best for a tofu box, and we feel no need to break with this tradition. If you feel you must put some coating on them, use a nontoxic salad bowl finish. The exterior of the tempeh incubator may be painted or lacquered, whichever you choose.

A few words about safety before you begin. Always use safety guards and eye protectors when operating power tools. Unplug the tool when changing blades and bits. When you get to the electrical installation instructions in chapter 5, exercise great care, and never have the unit plugged into a power source while doing any kind of work on the wiring.

Now it's time to get to work. As you turn the following pages you will see a great many photos and illustrations as well as instructions. They are all geared to help you build good-quality soybean equipment that will make you just a little more self-sufficient in the coming years by using soyfoods. Producing your own tofu and tempeh begins right here with the first pieces you will cut and assemble with the help of this book.

3 Tofu Box

The tofu box is fairly simple to build. It is a box with a removable bottom, three different-size tops, and adjustable baffles to make the box adaptable to three different capacities. You should have no trouble building it, if you follow the instructions. Take a look at blueprint sheets 3 and 4 before you begin to get a good idea of what each piece should look like and the position of each piece in the finished box. Blueprint sheet 3 details the pieces of the box, while blueprint sheet 4 details the pieces of the pressure system.

The labels listed on the blueprint sheets are used throughout this chapter. The numbered part of each piece's label corresponds to the number of the blueprint sheet where that piece is detailed, so you will find piece 3A, for example, on blueprint sheet 3. Before you begin building the pressure system and the box, decide which press to build. Your choice will affect the sizing of a few pieces. Chapter 1 describes both presses.

Chart 3-1 details the tools you will need to build your tofu box. These are only our recommendations; you may substitute as you wish. Chart 3-2 details the materials you will need to build your box, including the pressure plates to go with it. Do not deviate from this listing. You should note that we call for mahogany. For the use to which the box will be put, this is by far the best choice of wood. It is actually much easier to work with than many cheaper woods, and your box will last for many years. It is well worth the extra cost. Refer to chart 3-3 before you buy lumber, and be sure to get the right actual-size lumber when asking for a nominal size.

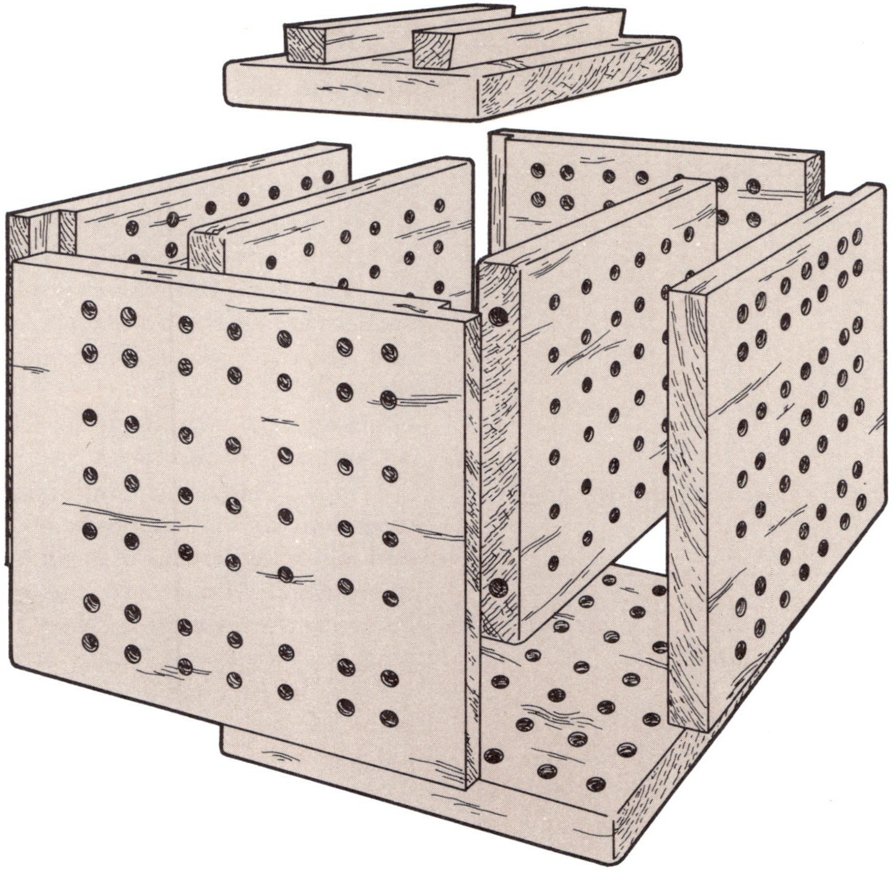

CUTTING

The first thing you will do to build the tofu box is cut the pieces. Then you will be able to follow through and assemble the box unhindered. The cutting diagram shown in illustration 3-1 is designed for the best economy of materials. The scrap left over from the cutting is shaded gray on the diagram, while the pieces to be used are in color. The rip cuts involved in this sequence of cuts are relatively short ones, making it easier for you to cut them. All of the pieces can be cut from a 6-foot-long 1 × 10 mahogany board, as shown

Tools

Circular Saw (with rip guide)
Handsaw or Saber Saw
Drill (with assorted bits)
Combination Square
Wood Chisel
2 Clamps
Screwdriver
Plane
Sandpaper
Wood Rasp (or other shaping device)
Tape Measure

Chart 3-1—These are the tools you will need to build the tofu box.

in the cutting diagram. A 1 × 10 should measure an actual size of about ¾ by 9¼ inches. Mark each cut and make the cuts one at a time, being sure the saw blade cuts on the side of the line away from the piece being cut. The direction of the wood grain on the pieces is important, so follow the cutting diagram, and you will eliminate structural problems later.

Before you make the first cut, be sure that the factory ends of the 1 × 10 are square. Then, starting at one end, measure 7¼ inches from the end to mark pieces 4C and 4B. Use a combination square to draw a line across the board at this point. (You will be marking all of your pieces in similar fashion, but mark and cut them one at a time.) Cut this piece off. From this section, you have to cut two pieces. Measure 3¾ inches from one 7¼-inch edge, and draw a line with the grain. Cut this piece off, and label it 4C. From the remainder of the section, mark and cut a piece 5¼ inches wide for piece 4B.

The next two pieces are squares. Cut straight across the 1 × 10, 7¼ inches from the end, for each piece. Then mark each piece for a cut 7¼ inches from the 7¼-inch factory edge, and cut with the grain. Label one piece 4A and the second piece 3A.

Now, instead of continuing in the same direction, go to the opposite end of the board, and follow the cutting diagram for the remaining pieces. Make sure the end is square before you start cutting. Next you will be ripping two strips the length of the remaining 1 × 10, the first one being cut at a bevel. This is the trickiest cut in the entire box. You cannot use a handsaw for this cut; the best way to do it is with a circular saw. A circular saw can be adjusted so the blade tilts at the proper angle, as shown in photo 3-1. Illustration 3-2 shows a cross-sectional detail of how the rip cuts are made. If you can't get these exact measurements out of your wood, they can be slightly smaller. You will be making one bevel cut to take the first strip from the board, then you'll make a straight 90-degree rip cut for

Materials

Description	Size	Amount
Lumber		
Mahogany	1 × 6	6 feet
Hardware		
#6 Flathead Brass Wood Screws	1″	16 (pot-mount press) 18 (table-mount press)
#8 Flathead Brass Wood Screws	1¼″	12
Cotter Pins	¼″ × 2½″	8

Chart 3-2—These are the materials you will need to build the tofu box.

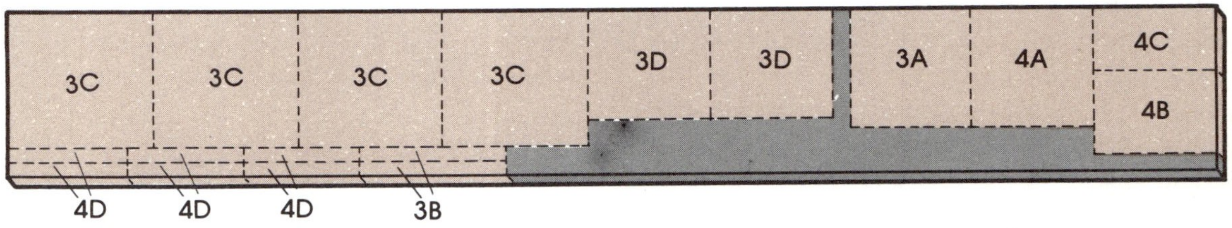

Illustration 3-1—The cutting diagram shows pieces to be cut from a 6-foot-long mahogany 1 × 10. Dimensions and labels are also shown. Scrap is denoted by the gray shaded area. Note that the pot-mount feet are not shown here, only the table-mount feet.

Nominal vs. Actual Lumber Size

Nominal Size	Actual Size
1 × 2	$\frac{3}{4}$" × 1$\frac{1}{2}$"
1 × 3	$\frac{3}{4}$" × 2$\frac{1}{2}$"
1 × 8	$\frac{3}{4}$" × 7$\frac{1}{4}$"
1 × 10	$\frac{3}{4}$" × 9$\frac{1}{4}$"
2 × 4	1$\frac{1}{2}$" × 3$\frac{1}{2}$"

Chart 3-3—Order your wood by the nominal size. If you cannot get a size, have another size cut down to the actual size.

the second strip. This will leave you with two identical strips, each with a 15-degree beveled edge, as shown.

Adjust the blade on your circular saw to a 15-degree angle first. Then install the rip guide on your saw so that it extends from the wide part of the base of the saw, as shown in photo 3-1. You will be cutting a strip with a beveled edge, making the top face of the strip $\frac{5}{8}$ inch wide and the bottom face $\frac{13}{16}$ inch wide. The rip guide will run along the edge of the 1 × 10 that is farthest away from the cut, as shown in photo 3-1. Mark the 1 × 10 on one end $\frac{5}{8}$ inch in from the long edge. Set your saw up with the front of the base resting on the end of the 1 × 10 and the outer face of the saw blade on the mark. The blade will tilt inward toward the wider part of the 1 × 10. Holding the saw in place, extend the rip guide until it fits the opposite edge.

When you have made the proper adjustments, you are ready to make the cut. Hold the saw steady and do not force it. Be careful to keep the rip guide against the edge while cutting. This will give you a nice, even cut. This is a potentially dangerous operation, and it requires care. Wear safety glasses to protect your eyes.

You will end up with a strip of wood that has one straight edge and one beveled edge. Now cut the next strip to match this one, as shown in illustration 3-2. This is a straight cut, requiring no bevel. READJUST YOUR SAW BLADE. Measure $\frac{13}{16}$ inch from the tip of the bevel on the wide face of the board, and mark the end of the board. Readjust the rip guide so that the outer edge of the saw blade meets this point. Now cut another strip of wood from the board. You will now have two identical strips from which you will cut the plunger guides and the feet.

With a handsaw or saber saw, cut six pieces from the strips, each measuring 6$\frac{3}{4}$ inches. Mark and cut them one at a time. Label each of these pieces 4D. The next two pieces to be cut from the strips are the feet. The size of these pieces will depend on whether you are going to make a table-mount or pot-mount press. If you are building the table-mount press, cut the feet 8$\frac{3}{4}$ inches long, as shown in illustration 3-1. If you are building the pot-mount press, cut the feet 5$\frac{3}{4}$ inches long instead of the length shown on the cutting diagram. Label both feet 3B.

Now go back to the wide board, and cut four pieces, each 8$\frac{1}{2}$ inches long. These should be labeled 3C.

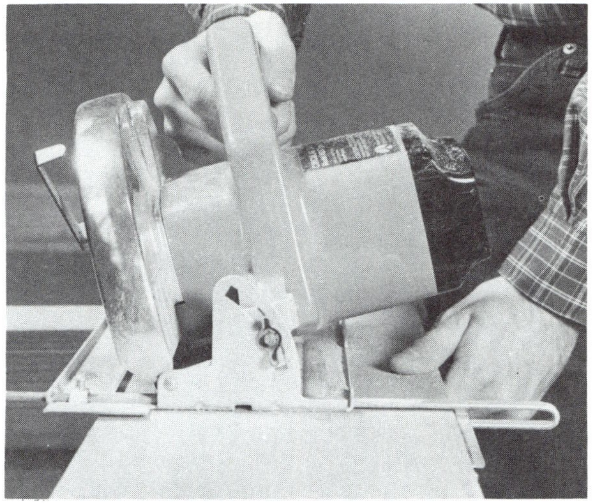

Photo 3-1—The bevel cuts for the pieces labeled 4D and 3B are done with the circular saw blade tilted 15 degrees. Note the use of the rip guide to keep the cut straight.

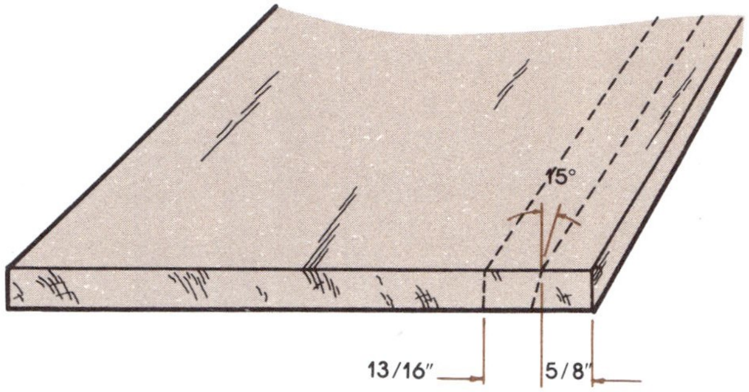

Illustration 3-2—Two strips are ripped from the 1 × 10; the first cut is a 15-degree bevel cut, and the second cut is straight. Note that one bevel cut gives you a bevel on both strips.

The remaining piece will be cut to form the two baffles. First you will mark a rip cut along the length of the remaining piece. Rip the piece down to a width of 6¾ inches. When you have the piece ripped, cut two pieces, each 7¼ inches long. Label the two pieces 3D.

This concludes cutting the pieces for the tofu box. There are some rabbet joints remaining to be cut, so keep your circular saw handy and proceed to the next phase of the construction.

RABBET THE SIDES

Next, you have to cut rabbets on the four side pieces labeled 3C. In a rabbet joint, one piece is notched on the end, and this notch

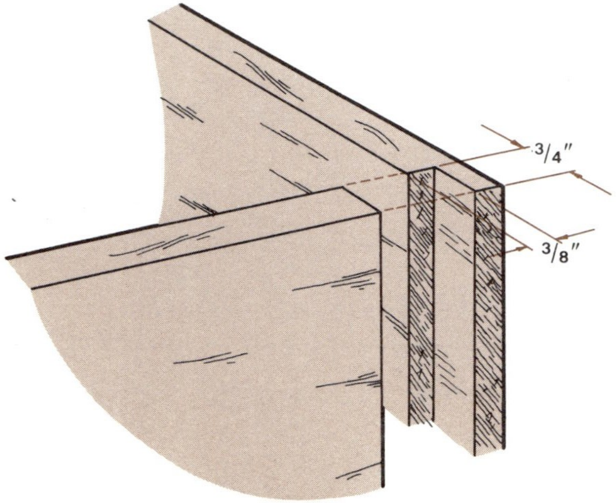

Illustration 3-3—In a rabbet joint, one side is rabbeted on one edge to fit onto the butt end of another side.

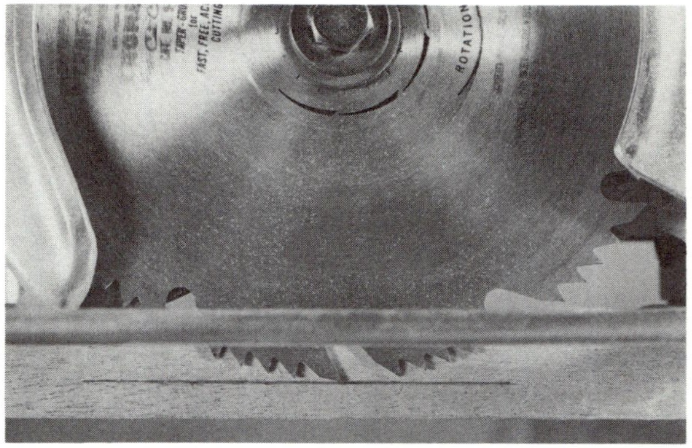

Photo 3-2—To set the depth of the rabbet cuts, draw a line on the edge of the piece to the desired depth, 3/8 inch, and raise or lower the saw blade until the lowest tip of the blade meets the line.

fits over the butt end of the piece being joined to it, as shown in illustration 3-3. You will be rabbeting only one end of each piece, as shown on blueprint sheet 3. Be sure you rabbet a 7½-inch end. These joints will add strength and stability to the finished box.

If you do not have a router, you can rabbet the sides with a circular saw by setting the blade depth to exactly half the thickness of the side piece, as shown in photo 3-2. With a ¾-inch thickness, this centerline will be ⅜ inch from the top. Be sure your saw is unplugged before doing this. Rest the base of the saw on top of the piece with the blade overhanging the end, as shown, and adjust the blade until it is exactly at the line. Try a test cut on a piece of scrap to be sure the depth setting is correct.

You should clamp the piece down so it will remain steady, and then clamp a straightedge to the piece to guide the saw while you cut. The best way to do this is to clamp both the straightedge and the side all in one operation, as shown in photo 3-3. You can use a piece of scrap wood as a guide, as long as it

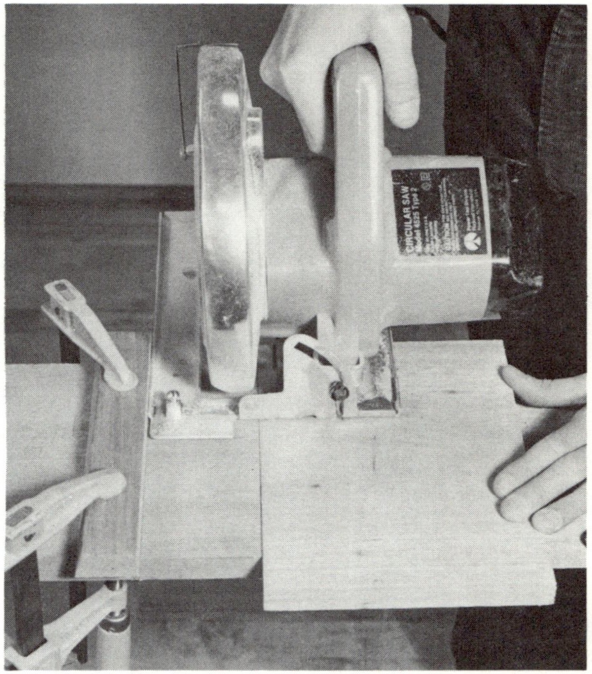

Photo 3-3—The cuts to form the rabbets on the side pieces are made using a circular saw with the blade adjusted to the correct depth. Note the use of a piece of wood as a straightedge clamped to the side piece and the work surface, and note the piece of scrap positioned next to the side piece to support the base of the saw.

has a straight edge. Do not try to cut the rabbets without clamping the pieces.

Look at photo 3-3 to see where the saw blade is in relation to the straightedge and how the saw rests on the piece being rabbeted. You'll have to place a piece of scrap of the same thickness or one of the other side pieces next to the end to be rabbeted, as shown. The bulk of the saw will then rest on this piece while you cut.

Keeping the base of the saw against the straightedge, mark the first cut, which will be ¾ inch in from the end. After this cut, continue to move the saw across the piece until all wood in the rabbet area is removed. Repeat this procedure on one edge of each of the four side pieces. Be sure your rabbets are uniform, or your box won't be square.

DRILL STRAINER HOLES

The holes through which the soymilk and whey will drain when the finished box is being used are drilled in a grid pattern. This turns the tofu box into a kind of strainer. The best way to mark the positions for the holes is to draw light intersecting lines on the face of the pieces and use the points where these lines intersect as center points for the holes. Follow the illustrations and refer to blueprint sheet 3 to clarify the hole positions. When drilling the holes, back the piece you are drilling with a piece of scrap wood so the drill goes directly into the scrap from the piece you are drilling. This will prevent the drill from splintering the other side of the piece and will give you a much neater box when you are finished. Drill all of the holes all the way through the piece.

Start with the sides. Mark two different grids on the four side pieces, one for the 5/16-inch holes used for positioning the baffles inside the tofu box and another grid for the rest of the holes. The first grid is a simple one, with four intersecting points at each corner of the piece, as shown in illustration 3-4. You will draw eight lines, four parallel to the 7½-inch end and four parallel to the 8½-inch edge. To do this, first draw a line ¾ inch down from one 8½-inch edge. Draw another line 1½ inches from the same edge. Then draw a third line 6 inches from the edge and a fourth line 6¾ inches from the same edge. Now draw the perpendicular lines, parallel to the 7½-inch edge. Draw a line 1 inch from the end that is NOT rabbeted. Draw another line 1¾ inches from the same end, a third line 6⅜ inches from the end, and a fourth line 7⅛ inches from the end. Using a 5/16-inch bit, drill a hole through the side piece at each intersection of the lines. Repeat this process on the other three side pieces.

The second grid you will draw on the face of the sides is for the ¼-inch holes. You'll use the lines already drawn, but add more lines to finish this grid. You will draw a series of lines parallel to one edge, then another series of lines perpendicular to and intersecting the first series of lines, as shown in illustration 3-5.

Begin by drawing a line 2¾ inches from

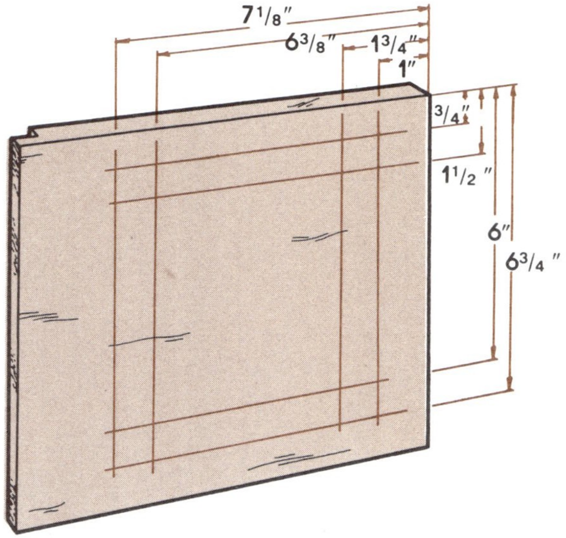

Illustration 3-4—To locate the holes in the side, eight lines are drawn for the 5/16-inch holes used for positioning the baffles.

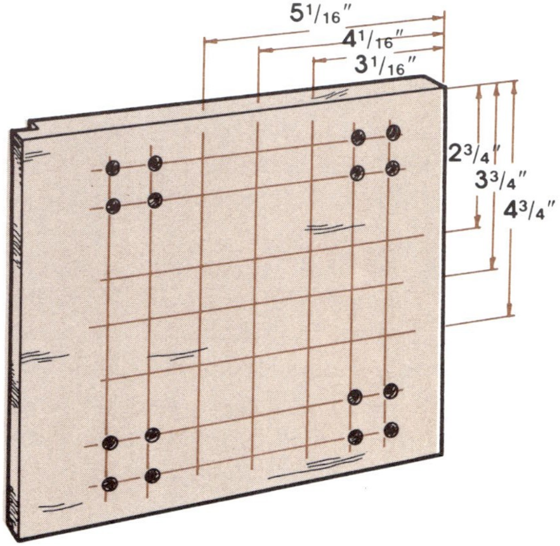

Illustration 3-5—The grid on the side piece is completed to locate the strainer holes that are to be drilled with a ¼-inch bit.

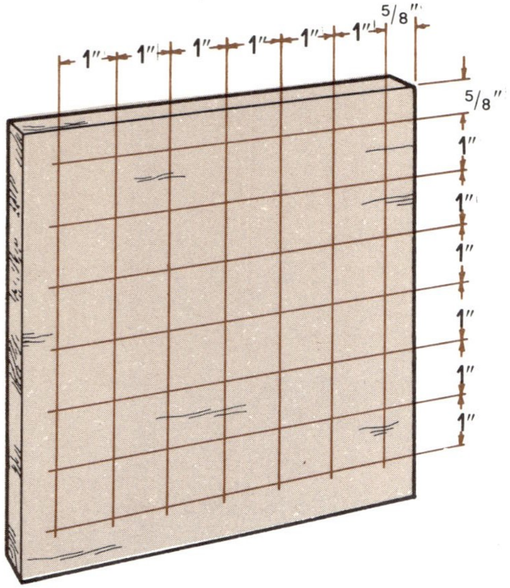

Illustration 3-6—The grid shows the location of the strainer holes in the bottom.

one 8½-inch edge, parallel to that edge. Draw another parallel line 3¾ inches from the same edge. The third line is drawn 4¾ inches from the edge.

Measure the next set of lines from the 7½-inch end that is NOT rabbeted. Draw the first line $3^{1}/_{16}$ inches from that end. Draw the second line $4^{1}/_{16}$ inches from the same end and the third line $5^{1}/_{16}$ inches from the end. Use a ¼-inch bit to drill holes in each of the points where lines intersect. Repeat this process on each of the sides.

The next grid you will mark is for the holes in the bottom, piece 3A. This is a simple grid forming 1-inch squares, as shown in illustration 3-6. You will be drawing a series

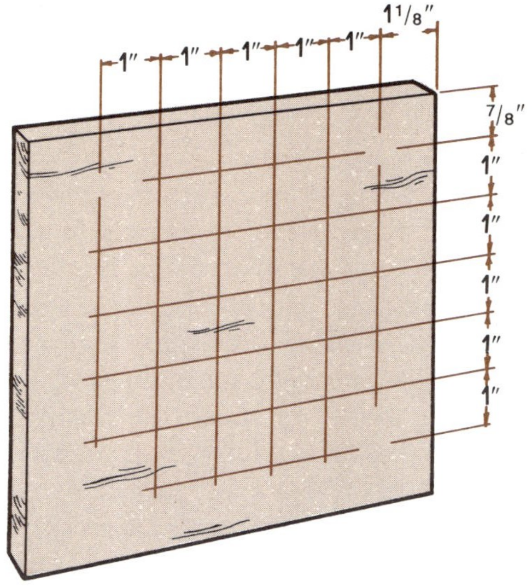

Illustration 3-7—The grid shows the location of the strainer holes in the baffles. Note that there are no holes drilled where the lines do not intersect.

of seven parallel lines and another series of lines perpendicular to and intersecting the first series of lines, as shown. Measure from one edge and work toward the opposite edge, drawing the first line ⅝ inch from the edge and a series of six more lines at 1-inch intervals. Repeat the process from the other end, drawing perpendicular lines. Drill a hole through the bottom at each intersection, using a ¼-inch bit.

Next, draw the grid pattern for hole positions in the two baffles, 3D. For each baffle you will draw six lines intersecting six more lines, forming 1-inch squares, as shown in illustration 3-7. Once again, you will work from one edge across the piece toward the opposite edge. Measure ⅞ inch from one 7¼-inch edge to draw the first line, and draw five additional lines at 1-inch intervals. Now measure 1⅛ inches from one 6¾-inch end, and draw a line parallel with the end, intersecting the first six lines. Then work your way across the piece, drawing lines 1 inch apart. Repeat the process for the other baffle. When you drill the ¼-inch holes through the baffles, DO NOT DRILL THE CORNER INTERSECTIONS. Look again at illustration 3-7 to be sure of which intersection points you should not drill.

ASSEMBLY

Now assemble the four sides. They are all the same size and will fit together in a square. Essentially, you are making a cube without a bottom and top. Look at blueprint sheet 3 for an overall view of how the sides join. In each of the four joints, a rabbet end of one side will fit over the butt end of another side. Taking each joint, one at a time, simply fit the butt end of one side into the rabbet of another side, keeping the top and bottom edges flush.

Drill a $^{5}/_{64}$-inch pilot hole through the rabbet into the end of the other side in three places. BE CAREFUL NOT TO DRILL THE PILOT HOLES DIRECTLY IN LINE WITH THE STRAINER HOLES IN THE FACE OF THE SIDE. The screws are long enough to block the holes. Before you fasten the joint, enlarge the holes through the rabbet with a $^{11}/_{64}$-inch bit for the shank of the screw. Countersink the hole on the outside face of the rabbet, using a countersink bit. Fit the two sides together again and fasten with 1¼-inch #8 flathead wood screws, making sure that the two pieces are square when fastened. The sides are not glued, just screwed. Join all four sides in a square in the same manner. With the sides assembled, you have only a few more details to finish before you have a completed tofu box.

Now is a good time to round the corners of the box and the other pieces to give a smoother, finished feel to the box. If you have a router, do this with a quarter-round bit. If you don't, then smooth the edges with a plane, a wood rasp, or other abrasive device. Sand all surfaces with fine sandpaper.

PLUNGER GUIDES AND FEET

Now you will attach two plunger guides, 4D, to each pressure plate. There are three different-size pressure plates, a large one to fit the entire box and two smaller sizes to accommodate the different baffle positions for changing the capacity of the box. When pressing, the plunger fits between the plunger guides on the pressure plate. The guides stabilize the plunger's position.

Take the large pressure plate, 4A, and fit two plunger guides to it, running across the grain of the pressure plate with the narrow face of the beveled plunger guide against the pressure plate, as shown in illustration 3-8 and on blueprint sheet 4. The guides are centered from edge to edge of the pressure plate, leaving ¼ inch from each end of the guide to the edge of the pressure plate. Draw a line running against the grain of the wood $2^{13}/_{16}$ inches from the end of the pressure plate, as shown in illustration 3-8. Draw another line $4^{7}/_{16}$ inches from the same end, parallel to the first line. The inside edge of the plunger guides should be positioned along these lines. The beveled edge of the guides will face outward, as shown, so they can be more easily gripped when you lift the pressure plate out of the box. Hold a plunger guide in position on one line with the narrow face against the pressure plate and, using a $^{1}/_{16}$-inch bit, drill two pilot holes through the guide and into the pressure plate, being careful not to drill all the way through the pressure plate. Widen the holes in the plunger guide for the screw shank with a $^{9}/_{64}$-inch bit, and countersink the holes. Fasten the plunger guide to the pressure plate with 1-inch #6 flathead wood screws. Repeat the same procedure, fastening a second plunger guide to the large pressure plate along the other positioning line.

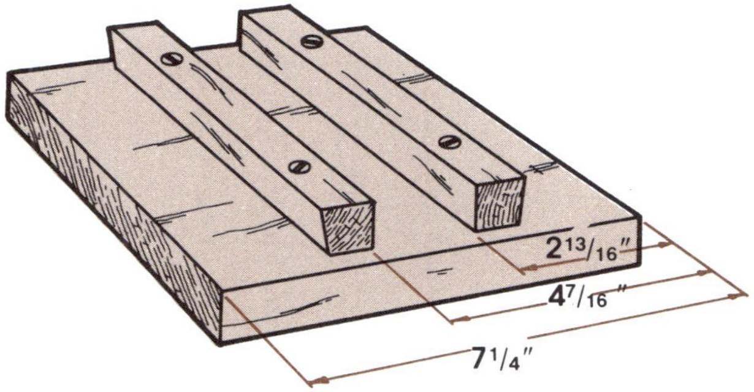

Illustration 3-8—The large pressure plate is shown with plunger guides positioned on it. Note the direction of the wood grain on the large pressure plate, as compared to that on the other two pressure plates.

Refer to blueprint sheet 4 and illustration 3-9 for the plunger guide positions on the medium-size pressure plate, 4B, and the small pressure plate, 4C. The positioning line on the medium-size plate will be $1^{13}/_{16}$ inches from the 7¼-inch edge, and the second line will be $3^{7}/_{16}$ inches from the same edge. The plunger guides on the small pressure plate are positioned in the same manner, but the first line will be $1^{1}/_{16}$ inches from the 7¼-inch edge, and the second line will be $2^{11}/_{16}$ inches from the same edge. Note that for both the small and the medium-size pressure plates, the guides will run with the grain.

You should still have two beveled pieces marked 3B left. These are the feet to be fastened to the bottom, 3A. You've already decided which press you will be building, and you've cut the feet to the proper length for that particular press. The feet for the pot-mount press are short so that they fit between the frame rails of the press and keep the tofu box from sliding from its position on the press. The feet for the table-mount press are longer so they protrude from the edges of the bottom. The box rests on them, and they keep the sides elevated from the tray.

If you are building the pot-mount press,

you need two feet, each measuring 5¾ inches. Working with one at a time, position them in the space between the first two rows of holes in the bottom, with the narrowest edge against the face of the bottom and the bevel to the outside, as shown in illustration 3-10. BE SURE THE FEET GO ACROSS THE GRAIN, as shown. Each piece should be set back from the edge ¾ inch. To attach the feet, hold one in place, making sure that it does not block any holes, and drill two $\frac{1}{16}$-inch pilot holes through the foot and into the bottom. Be careful not to drill all the way through the bottom. Using a $\frac{9}{16}$-inch bit, enlarge the holes in the feet only, and countersink the holes. Fasten each foot with 1-inch #6 flathead wood screws. Repeat this process to fasten the other foot.

The feet for the table-mount press, each 8¾ inches long, are positioned in the same manner except that they overhang the edge of the bottom by ¾ inch on each end, as shown in illustration 3-11. Note that each of these pieces are fastened with three screws.

INSTALL THE BAFFLES

There is one final thing to do to complete your tofu box. The box has two baffles, 3D, to allow you to change the interior size of the box to suit the quantity of tofu you want to make. These are held in place inside the box by cotter pins slipped through holes in the side of the box and then into holes in the end of the baffle. Each baffle needs two holes on each 6¾-inch end, as shown in illustration 3-12.

To mark the hole positions on the baffles, first set up the box, placing it over the bottom. It should rest on the protruding feet if you've built the table-mount press, and if you have the pot-mount press, you'll have to use scrap pieces to set underneath the box to act as extensions on the feet so that the box is

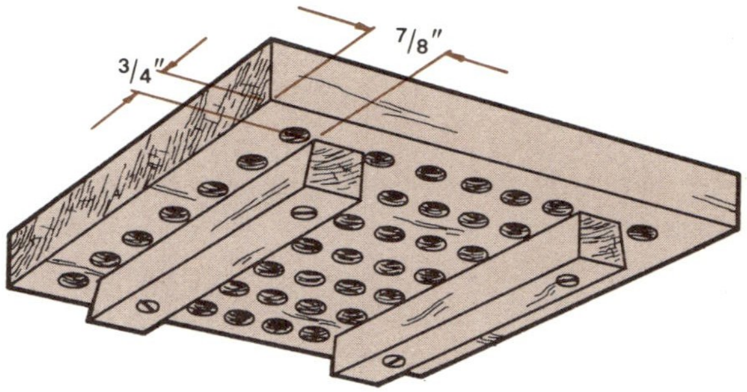

Illustration 3-10—The feet for the pot-mount press are positioned on the bottom as shown. Note that the feet do not block the strainer holes.

Photo 3-4—Holes for the baffle positioning are marked with a pencil through the $\frac{5}{16}$-inch holes in the box onto the edge of the baffle. Note that for the pot-mount press tofu box, shown here, pieces of scrap have been placed underneath the box to elevate it to the correct height.

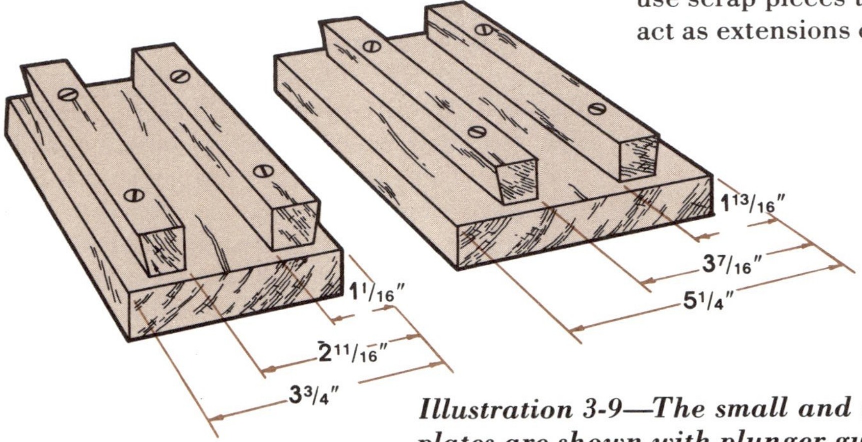

Illustration 3-9—The small and medium-size pressure plates are shown with plunger guides in position. Note the direction of the wood grain on the pressure plates.

elevated ¾ inch. The bottom should be flush with the lower edges of the sides. When this is set up, take a baffle, and place it inside the box, flat against one side. It doesn't matter which side, but make sure you have the grain on the baffle running horizontally, not vertically. In this position the end of the baffle should be centered over the first row of holes in the side. If not, you may have to move the baffle a little to allow for this. Hold the baffle in position and mark the end of the baffle through the uppermost $^5/_{16}$-inch hole, as shown in photo 3-4. Mark the baffle the same way through the second $^5/_{16}$-inch hole up from the bottom. Now take the baffle out of the box and check that your pencil marks are centered on its end. Drill ¼-inch holes, 1½ inches deep, into the end of the baffle. Use the box as a drilling jig to be sure of alignment. Drill holes in the other end of this baffle in the same manner on the opposite side of the box. You should be able to fit a ¼ by 2½-inch cotter pin through the side of the box and into the corresponding hole in the baffle in each baffle position. Two cotter pins in each side of each baffle will hold the baffle in place. Repeat this process with the remaining baffle on the opposite side of the box. If you have drilled the holes accurately, the pin holes on the baffles will match holes on any of the four sides. This completes the tofu box assembly.

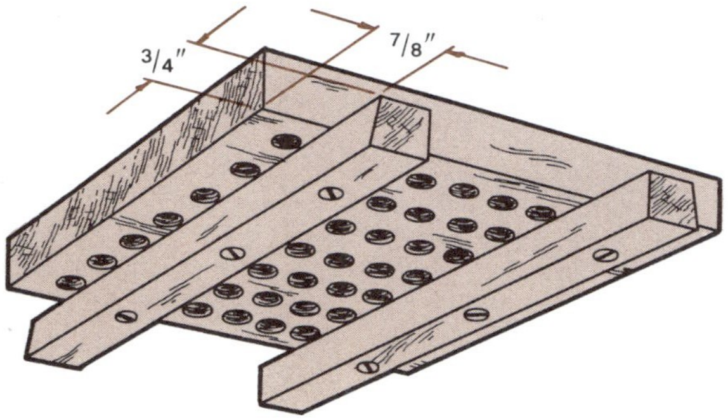

Illustration 3-11—The feet for the table-mount press are positioned on the bottom as shown. Note that the feet do not block the strainer holes.

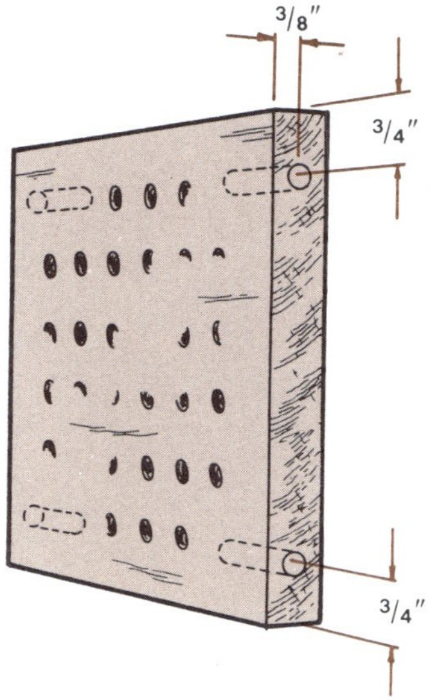

Illustration 3-12—The four holes in the ends of the baffle for mounting it inside the box are shown. Note the direction of the wood grain on the baffle in relation to the holes.

4 Pot-Mount Press

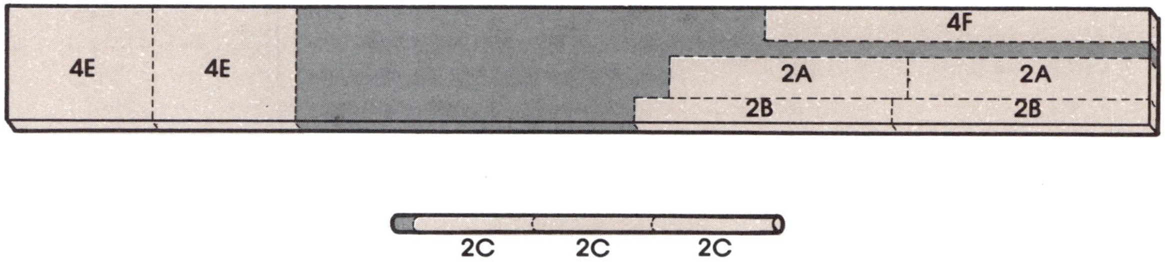

Illustration 4-1—The cutting diagram shows the pieces to be cut for building the pot-mount press, with the dimensions and label of each piece. Gray shaded area denotes scrap.

The pot-mount press will serve as a base for the tofu box and produce the pressure needed for pressing the soymilk and tofu. It is a derricklike structure in appearance. Its foldable arms extend upright to provide a pivot point for the lever to apply pressure, and they fold down against the frame of the unit for easy storage.

Chart 4-1 details the tools you will need to build a pot-mount press. In designing the press, we wanted a unit that could be built without the aid of stationary power tools. As you can see, we were successful. However, if you have stationary power tools, by all means use them. Chart 4-2 lists all the materials you will need to build the press.

Tools

- Handsaw
- Saber Saw or Coping Saw
- Hacksaw
- Drill (with assorted bits)
- Hammer
- Combination Square
- 2 Clamps
- Screwdriver
- Plane
- Sandpaper
- Wood Rasp (or other shaping device)
- Round File (or other shaping device)
- Tape Measure
- Wrench (for 1/4-20 nut)

Chart 4-1—These are the tools you will need to build the pot-mount press.

CUTTING

The pot-mount press has only a few parts, so the cutting phase will not have many steps. The steps in the cutting sequence may be few but they can be tedious. Rip cuts can be difficult, but with a good sharp handsaw or a saber saw, you'll be able to make the cuts safely and smoothly. We do not recommend that you use a circular saw; the pieces are too small to be cut safely with a saw that large.

Working with the cutting diagram in illustration 4-1, mark and cut each piece one at a time. Be sure the saw blade cuts on the side of the line away from the piece to be cut. You'll be making all of your pieces from a single 6-foot-long 1 × 8 mahogany board. Check that the ends of the board are square before you begin. If they are not, square them. The first two cuts are simple. From one end of the 1 × 8, measure 9 inches and draw a straight line across the board, using a combination square, and cut the piece off. The next piece is also cut 9 inches from the end, giving you a piece identical to the first. Label them both 4E. These are the two plunger halves, which will be assembed and shaped later.

The next cut will be a rip cut. To mark the cut, draw a straight line at least 32½ inches long lengthwise on the board, 1¼ inches from the edge. You only have to cut far enough to get two 16-inch pieces. Hold the board steady or clamp it, and cut the 1¼-inch strip. To save time and materials, only cut as far as you need. Cut two pieces, each 16 inches long, and label both pieces 2B.

These are the arms of the press, and they must be tapered at one end, as shown on blueprint sheet 2. Take one piece labeled 2B, and lay it on its wide face. Measure 3 inches from one end, and draw a straight line across the piece. Then draw a line on the end of the piece, ⅜ inch from the face. On the edge of the piece, extend the two lines at an angle so they connect with each other, forming a wedge, as shown in illustration 4-2. To cut this wedge off, clamp the piece down and use a handsaw or saber saw. Repeat this process

Materials

Description	Size	Amount
Lumber		
Mahogany	1 × 8	6 feet
Hardwood Dowel	⅜″ diam.	2 feet
	⅛″ diam.	1 foot
Hardware		
¼-20 Carriage Bolt	3″	1
Flat Brass Washers	¼″	3
¼-20 Nut	. . .	1
Copper Tubing	⅜″ OD	⅞ inch
3/32″ Hitch Pin	1⅝″	1
¼″ Clevis Pin	2″	1
Cup Hook	1″	1
Glue		
Water-Resistant Wood Glue	. . .	½ pint

Chart 4-2—These are the materials you will need to build the pot-mount press.

with the other arm, first marking it, then cutting the taper. Put the arms aside for now, and go to the next cut on the 1 × 8.

The next two pieces are ripped from the board the same way you ripped the pieces for the arms. Measure 2⅜ inches from the edge, and draw a straight line lengthwise on the board for a distance of at least 30½ inches. Make a rip cut along this line for that distance, then cut off two 15-inch pieces from the strip. Hold the strip firmly while cutting. Label both pieces 2A. They will be notched later after the press is assembled.

The final rip cut to be done is shorter, but by now you have a narrow piece of wood to work with, so if you are not clamping the piece, make sure that you hold it steady while cutting. Measure 1⅞ inches from the edge, and draw a straight line lengthwise for at least 23 inches. Make a rip cut along this line. Cut this strip off at the 23-inch mark, and label it 4F. This is the lever; it will be contoured later in the assembly phase.

The only pieces left to be cut for the pot-mount press are the three dowels. From a length of ⅜-inch-diameter hardwood dowel,

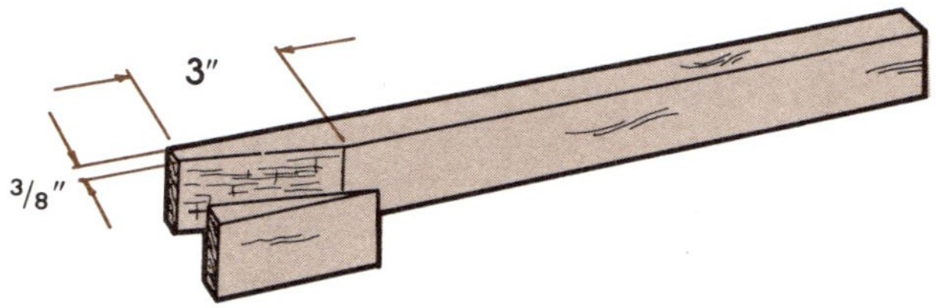

Illustration 4-2—The arm is tapered on the end where it fits against the frame rail.

cut three sections, each measuring 7⅜ inches, and label them 2C. Mark and cut them one at a time to allow for the thickness of the saw blade. The length is critical; it should be exactly 7⅜ inches. To cut the dowel requires a light touch.

ASSEMBLY

You now have a neat little pile of pieces. The first thing you'll be concerned with is the frame assembly, which includes the two frame rails, 2A, and two of the dowel pieces, 2C. Pull these pieces from the pile and check their positions on blueprint sheet 2. The frame is comprised of two frame rails lined up parallel to each other and connected by a dowel at each end. The dowels are pegged in place with a ⅛-inch-diameter dowel, 2D, as shown in photo 4-1 and on the blueprint. The ends of the dowel will be flush with the outer face of the frame rails.

Before you begin to mark the center points for the dowel holes, mark one edge of each frame rail as the upper edge. You will refer to this edge throughout the assembly. To mark the hole position for the first ⅜-inch-diameter dowel, measure 1⅜ inches from one end of the rail and ½ inch from the upper edge, as shown in illustration 4-3. Mark this as the center point for the hole. Do the same on the other frame rail. Using a pencil, label these center points with the letter *A* to distinguish them from the other holes you will be drilling. On the same end of the frame rails, mark the center point for another dowel hole by measuring 2¼ inches from the end and 1⅜ inches from the upper edge, as shown. Do the same on the other frame and mark both center points *B*.

Go to the opposite end of the frame rails, and mark the center point for the third dowel.

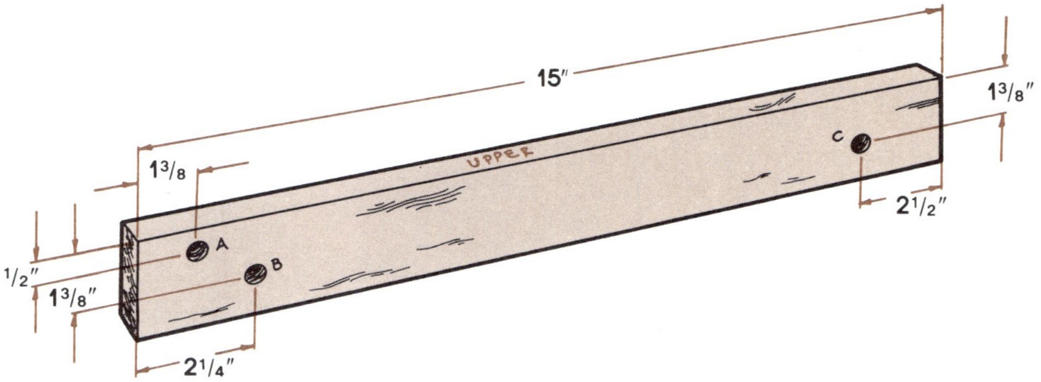

Illustration 4-3—There are three hole positions for dowel insertions on each frame rail. Note that they are labeled* A, B, *and* C, *and that the upper edge is labeled also.

Measure 2½ inches from the end and 1⅜ inches from the upper edge, as shown in illustration 4-3. Mark the same point on the other frame rail. These two holes get labeled *C*. You should now have six center points: two marked *A*, two *B*, and two *C*. Drill a hole through the frame rail on each of the six points, using a ⅜-inch bit.

The next step is to insert the dowels to form the frame. To assemble the frame, you will insert one end of a dowel into the hole marked *A* in one frame rail and the opposite end of the dowel into the same hole position in the other frame rail. Connect the holes marked *C* in the same manner. The holes marked *B* will be doweled when you install the arms; do nothing with them now. With a light hammer, tap the dowels into one frame rail until they are flush. Then with a flat, hard surface to back up the dowel ends already inserted, tap the other frame rail onto the dowel ends that are sticking up until they are flush with the surface of the second frame rail. Make sure you are fitting the dowels into the correct holes. Take another look at blueprint sheet 2 and illustration 4-3 to make sure of the positions.

Next, you have to fasten the dowels in their positions by pegging them with a small section of ⅛-inch-diameter hardwood dowel. The dowel peg, 2D, is tapped into a pilot hole through the upper edge of the frame rail and into the ⅜-inch-diameter dowel, as shown in photo 4-1. This will lock the dowels in place.

First, make sure that the dowels are still flush with the outer face of the frame rails before drilling pilot holes for the pegs. All of the holes will be drilled from the upper edge of the frame rails. Using a ⅛-inch bit, drill a hole into the upper edge of the frame rail and through the ⅜-inch-diameter dowel inserted in the frame rail. Do not drill the hole all the

Photo 4-1—A small section of 1/8-inch dowel tapped into the frame rail and into the 3/8-inch dowel will hold the frame assembly together. The same method is used to fasten the arms to the 3/8-inch pivoting dowel.

way through to the other edge of the frame rail, just through the dowel. Take a small section of ⅛-inch-diameter dowel and tap it into the frame rail and into the dowel. This will hold the ⅜-inch-diameter dowel in position and strengthen the joint. With a fine-tooth saw such as a hacksaw, cut off the dowel peg flush with the upper edge of the frame rail. The ⅜-inch-diameter dowel will now be locked in position. Repeat this process with the other three joints, making sure the ⅜-inch-diameter dowels are flush with the frame rails before pegging them. With fine sandpaper, sand over the ends of the pegs to make them even with the surface of the frame.

ARM ASSEMBLY

Next, install the arms on the frame. The arms are designed to fold into the frame assembly for easy storage. When in use they are in an upright position. The arms are joined to the frame by means of a ⅜-inch-diameter dowel through both frame rails and arms.

The holes for this dowel position are already drilled in the frame rails. Now, drill holes for the pivoting dowel in the tapered ends of the arms. Mark the center point for the holes as shown in illustration 4-4. Working on the flat face, mark a center point ⅝ inch from the edge of the arm and 2 inches from the end. Be sure to mark on the flat face of the arm, not on the taper.

You may get confused when you try to drill the holes in the arms. The hole you drill has to be perpendicular to the tapered face of the arm, NOT perpendicular to the flat face, although you've marked the center point on

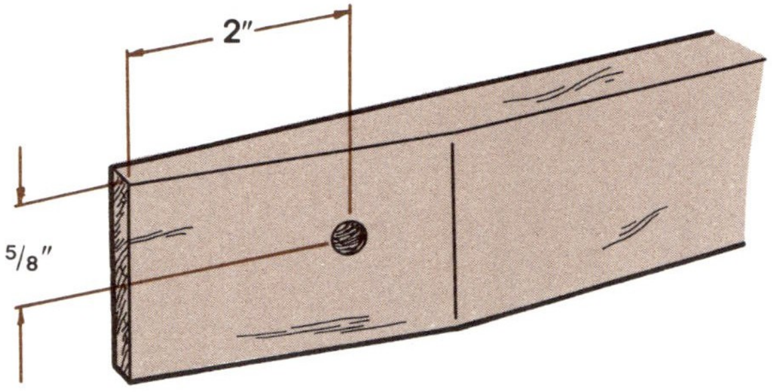

Illustration 4-4—The dowel hole in the arm is positioned on the taper. The hole will be drilled from the opposite face, not from the tapered face.

the flat face. To be sure that you are drilling a straight hole in relation to the holes in the frame rails, hold the tapered face of the arm against a piece of scrap wood and drill straight down into the scrap, using a ⅜-inch bit, as shown in photo 4-2.

When you fit the tapered face of the arms against the inside face of the frame rails, you should be able to match up the holes and tap a ⅜-inch-diameter dowel through all four holes. The arms will angle toward each other at the top when they are properly positioned. When you fit the arms to the frame, be sure that the tapered ends of the arms are under the dowel in hole *A* and that the opposite ends of the arms rest on top of the dowel in hole *C*.

Next, peg the dowel in position with ⅛-inch-diameter dowels IN THE ARMS ONLY. Make sure that the dowel ends are flush with the outer faces of the frame rails and that the taper of each arm fits flat against the frame rail. Then, using a ⅛-inch bit, drill a hole into the upper edge of the arms and through the ⅜-inch-diameter dowel. The peg hole should not go all the way through to the other edge of the arm, just through the dowel. Cut a small section of ⅛-inch-diameter hardwood dowel long enough to reach through the ⅜-inch-diameter dowel, and tap it into the hole to lock the arms in place. Cut off any part of the dowel peg sticking out of the edge of the arm. Do this to both arms.

The final step in the arm assembly is to install the copper tubing and the carriage bolt, connecting the tops of both arms. The copper tubing serves as a spacer between the

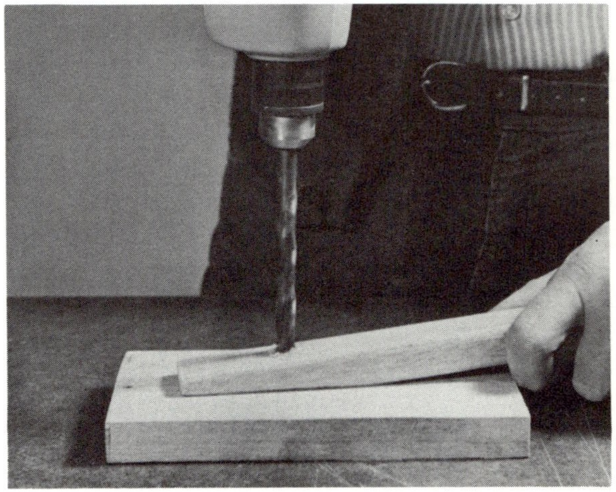

Photo 4-2—The dowel hole is drilled through the arm by holding the taper of the arm flat against a piece of scrap wood, keeping the drill perpendicular to the piece of scrap.

arms and as a catch for the notch in the lever.

The holes for the carriage bolt must be aligned with each other so that the bolt can pass through both arms at an angle. To draw guidelines on the edge of the arms to help you drill the holes at the proper angle, mark a point ¾ inch from the top end of each arm. Hold a straightedge on these marks across the top edge of both arms, and draw a line across both edges. You should end up with lines as shown in photo 4-3. Take a good look at photo 4-3, and drill a 5/16-inch hole through both arms, lining up your drill with the guidelines, as shown. Hold each arm firmly as you drill and keep your drill straight.

Next, using a hacksaw, cut a piece of ⅜-inch (outside diameter) copper tubing ⅞ inch long. This does not have to be angled; just cut it off straight. Now insert a ¼-20 carriage bolt 3 inches long through one arm, then slip a ¼-inch flat washer onto it and pass it through the ⅞-inch section of copper tubing and slip another 1/4-inch flat washer onto it. Then push the bolt through the other arm and fasten it with another ¼-inch flat washer and a ¼-20 nut, as shown in photo 4-4, and tighten it with a wrench. Do not tighten the bolt too much or the copper tubing will begin to buckle or bend under the stress.

NOTCHING THE FRAME RAILS

With the entire frame and arm assembly completed, you can notch the lower edges of the frame rails to fit the particular pot you have chosen to use with the unit. The notches keep the press from slipping off the pot while making tofu.

The inside diameter of the pot you choose may range from 13 inches to 14½ inches. You may be tempted to use a smaller pot than the minimum size stated here, but a smaller pot will not be wide enough to catch all of the liquid when using the press. We've found that canning kettles and clam steamers are ideally suited for this purpose.

To mark the lines for the notches, place the press across the top of the pot with the arms extended in the upright position, as shown in photo 4-5. The lower, tapered ends of the arms should extend below the frame rails. These should be against the inside of the pot, and the press should be centered from side to side.

Now trace the inside curve of the pot onto the lower edge of the frame rails at all four points where the rails cross the brim of the pot, as shown in photo 4-5. The frame should not fit the pot tightly. It is better if it has some play. To give the right amount of play, cut on the inside of the traced lines when making the notches. Cut on the inside of each line, into the frame rail, to a depth of ½ inch. A handsaw is recommended for this.

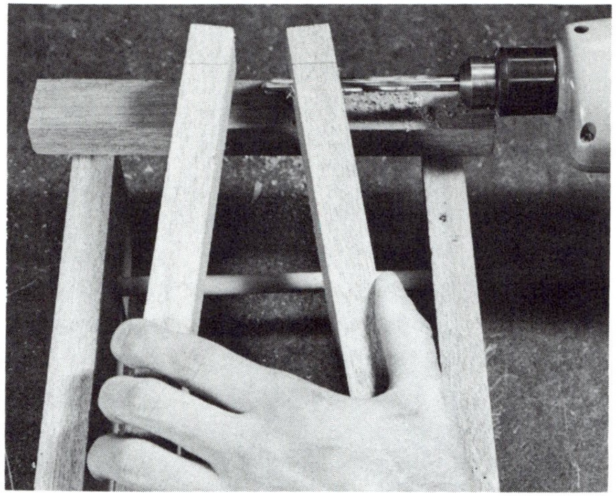

Photo 4-3—Drill a clearance hole for the carriage bolt through the top of both arms, keeping the drill aligned with the guidelines drawn on the edge of the arms, as shown. Note that the arms are supported by a piece of scrap wood placed across the frame.

Photo 4-4—The clearance holes at the top of the arms are fitted with a carriage bolt which is pushed through a washer, a small section of copper tubing, and another washer and fastened with a washer and nut on the outside face of one arm.

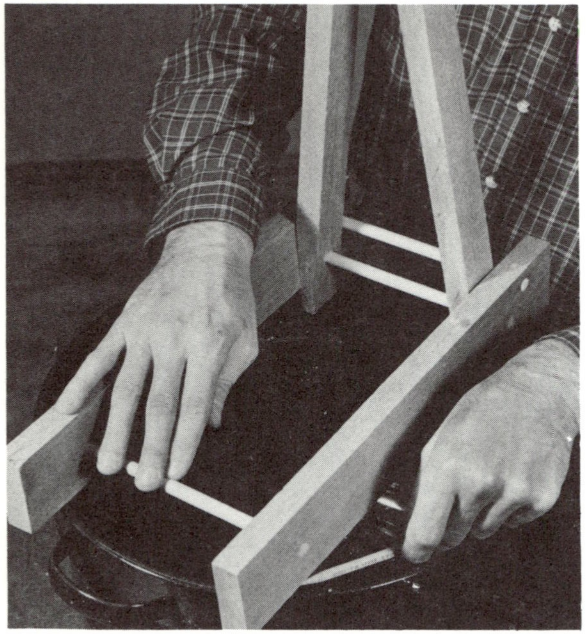

Photo 4-5—The press is positioned across the center of the pot when tracing the inside of the pot brim for notching the frame. Note that the arms are upright and that the bottom ends of the arms are inside the pot.

Now draw a line from the depth of the cut to the end of the frame rail, parallel with the edge, on both faces of each frame rail, and extend the lines across the end. Complete the notch by cutting along these lines to meet the first cut, as shown in photo 4-6. Repeat this procedure to notch both ends of each frame rail.

You are now finished with the most complicated parts of the press, but there are still two very important units to be dealt with: the lever and the plunger. Refer to blueprint sheet 4 for a detailed view of the lever and plunger.

SHAPING THE LEVER

Piece 4F is the lever. The lever is contoured to give the end of the lever a better grip and to provide a more aesthetic look. Illustration 4-5 details the cuts, which can be made with a saber saw or coping saw. Round the edges after cutting. Measure 1½ inches from one edge, and draw a straight line parallel to the edge for a distance of 4 inches from the end of the piece, as shown. Angle this line out to the edge at a point 5½ inches from the end. Cut off this section to form the hand grip of the lever. Round all edges of the lever until it feels comfortable.

A hole in the lever, for hanging the plunger, is positioned 6⅜ inches from the square end of the lever and ½ inch up from the lower edge, the edge that has been cut to form a hand grip. Use this as the center point for drilling a ¼-inch hole in the lever.

Next, notch the top edge on the end opposite the hand grip. This is done by first drilling a hole. Mark a point 1 inch from the end and ½ inch from the edge. Using a ½-inch bit, drill a hole with this mark as the center point. Once the hole is drilled, make two small cuts from the edge of the lever to each side of the hole. This will open the hole to the edge, forming a notch ½ inch wide, as shown. Last, install a 1-inch cup hook in the opposite end of the lever in the butt end of the hand grip. Use a $^{1}/_{16}$-inch bit to drill a pilot hole for the cup hook. Round all edges and ends to give the handle a nice look and feel.

SHAPING THE PLUNGER

The plunger is a somewhat tricky item to build. It has contours that give it a very aesthetic look. Look at blueprint sheet 4 to get a good idea of what you will be building. Then turn to the actual cutting template on page 65 for instructions on cutting out the plunger halves, pieces 4E.

Once the two halves are cut to their

Photo 4-6—The notches on the frame are made with two cuts. The first cut is on the line traced from the brim of the pot, and the second cut is from the end of the frame rail to the first cut.

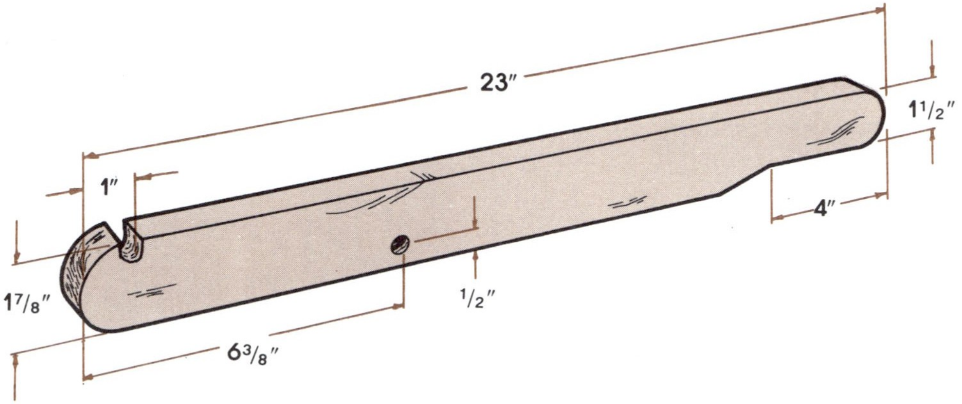

Illustration 4-5—The lever, as it is finally shaped, is narrowed on one end to form a handle grip and rounded on both ends. There is also a hole for hanging the plunger and a notch for hooking into the catch at the top of the press arms. The cup hook is not shown. Note that the measurements of the lever for a table-mount press are different.

Photo 4-7—Clamp the plunger half upright, as shown, to make the first cut for the notch in the plunger half.

proper shape, notch the stem of each half (the narrow part that is rounded on the top). The notch is cut out of each half, as shown on the blueprint, so that when they are glued together the two notches will form one notch wide enough to fit around the handle lever. First, mark *inside* on one face on each half. These are the faces that will be glued together. Then take one piece, and measure $^{5}/_{16}$ inch from the outside face on the edge of the stem. Make several marks along the edge of the stem, up around the curved top and along the opposite edge. Connect these marks with a straight line on the edge of the piece, ending 2⅝ inches from the top of the curves on both edges. This line can be drawn with a steady hand, running your pencil along the edge with your fingers in a fixed position. Now draw a straight line 3 inches from the top across the inside face of the stem. Extend both ends of this line around the corners, and angle it to meet the end of the line on the edge. Now you have the entire shape of the notch drawn. With a fine-tooth saw, cut along the lines to remove the wood that forms the notch. You'll have to clamp the piece upright when cutting into the edge, as shown in photo 4-7. Do not try to hold it with your hand. Cut to the end of the line and no further. Then cut on the second line at an angle from the inside face to meet the end of the first cut. The notch will angle inward when the two halves are glued together. This angle will give the stem of the plunger added strength and prevent splitting. You can curve the inside corner of the notch with a round file to make the angle more graceful. Repeat the procedure on the other piece labeled 4E to form the other half of the notch.

The next step in building the plunger is to put the two halves together. Glue them with the edges flush and the notches facing each other, forming a single notch wide enough that the ¾-inch thickness of the lever will fit in loosely. We recommend using a water-resistant wood glue for this job. Clamp the two halves together after gluing, and give them a chance to set. Follow the directions given on the glue container for its application. After the glue is set, drill the holes from which the plunger will hang in the stems near the top. Use a ¼-inch bit to drill the hole in the center of each stem, ¾ inch from the top. When in use, the plunger will hang from the lever by means of a ¼-inch clevis pin, 2 inches long, inserted into the hole in one stem, through the hole in the lever, and through the hole in the other stem, then fastened with a $^{3}/_{32}$-inch hitch pin, 1⅝ inches long.

This completes the pot-mount press.

5 Table-Mount Press

The table-mount press is a sturdy little bench with folding legs and arms which open to an upright position for use and fold into the deck of the press for storage. The deck is constructed in a manner that will remind you of a butcher block, but it is much easier to build.

The woods used were selected for their performance under moist conditions. Two factors dictated the use of two types of wood here. Mahogany is used, as it was in the tofu box and pot-mount press assemblies, for its general suitability for the project, but redwood was chosen for the deck pieces because of its extra durability and moisture resistance. It will lend more stability to the table-mount press.

Follow the photos and illustrations as you proceed through the chapter, and refer to blueprint sheets 1 and 4 for further clarification. Label each piece with a pencil as you cut it. The number of the label corresponds to the number of the blueprint sheet on which that piece is detailed. You will find piece 1A, for example, on blueprint sheet 1. This feature helps you to turn to the blueprint sheets for quick reference. Before buying your lumber, take a look at chart 3-3 on page 41 to be sure you get the right actual size when you ask for nominal size lumber.

Chart 5-1 shows the tools you will need for this project. Of course, you should feel free to use any other tools you may have. The list is the minimum you need to complete the press. Chart 5-2 details the materials you will need to build your press. Follow the materials recommendations very closely, and be wary of substituting materials.

CUTTING

You will be cutting from two types of lumber for the table-mount press. First, you will cut from a 1 × 8 mahogany board, then you'll cut pieces from a redwood 2 × 4. There are a number of rip cuts to be done, and these can be tedious, but we've figured the rip cutting so that the length of these cuts is kept to a minimum. You will cut into the wood only as far as you have to in order to obtain the pieces you need. You don't have much space to hold onto the wood while cutting, so it is best to use a saber saw or a handsaw rather than a circular saw.

Tools

- Handsaw
- Saber Saw or Coping Saw
- Hacksaw
- Drill (with assorted bits)
- Ratchet Wrench (with 7/16-inch socket)
- Hammer
- Combination Square
- 2 Clamps
- Screwdriver
- Plane
- Sandpaper
- Wood Rasp (or other shaping device)
- Chisel
- Tape Measure

Chart 5-1—These are the tools you will need to build the table-mount press.

Follow the cutting diagram in illustration 5-1. Mark and make each cut before going on to the next. Otherwise, you will not be allowing for the thickness of the saw blade, and your pieces will not be the right size. Check that the ends of the lumber are square before you begin to cut, and square them up if they are not. You will label each piece with a letter and a number as you cut it. These labels will be referred to later in the text, and they will help you to check the pieces with the details on the blueprint sheets.

Cut the mahogany pieces first. You will be able to get all of the mahogany pieces from a 6-foot-long 1 × 8 board. Mark a line 9 inches from the end of the 1 × 8, first making sure that the end is square. Use a combination square to draw the line. Cut the piece off along this line. Repeat these steps and label both pieces 4E. These are the plunger halves, which will be contoured later with a saber saw. Put them aside for now, and go on to the next cut.

The next cut is a rip cut the entire length of the remainder of the board. You will rip a 2½-inch-wide strip from the board, and then cut this into smaller pieces, as shown in illustration 5-1. Mark a line 2½ inches from one edge for the length of the board. Using a saber

Materials

Description	Size	Amount
Lumber		
Mahogany	1 × 8	6 feet
Redwood	2 × 4	4 feet
Hardware		
#8 Roundhead Brass Wood Screws	2″	4
#8 Flathead Brass Wood Screws	$1^1/_2$″	8
Flat Brass Washers	$^1/_4$″	12
$^1/_4$-20 Flat-head Bolt	$2^1/_2$″	1
$^1/_4$-20 Threaded Rod	$7^1/_2$″	2
$^1/_4$-20 T-nut Fasteners	. . .	3
$^1/_4$-20 Hex-Head Nuts	. . .	2
Copper Tubing	$^3/_8$″ OD	$^{13}/_{16}$ inch
$^3/_{32}$″ Hitch Pin	$1^5/_8$″	1
$^1/_4$″ Clevis Pin	2″	1
Cup Hook	1″	1
Cookie Sheet	10″ × 15″	1
Glue		
Water-Resistant Wood Glue	. . .	$^1/_2$ pint

Chart 5-2—These are the materials you will need to build the table-mount press.

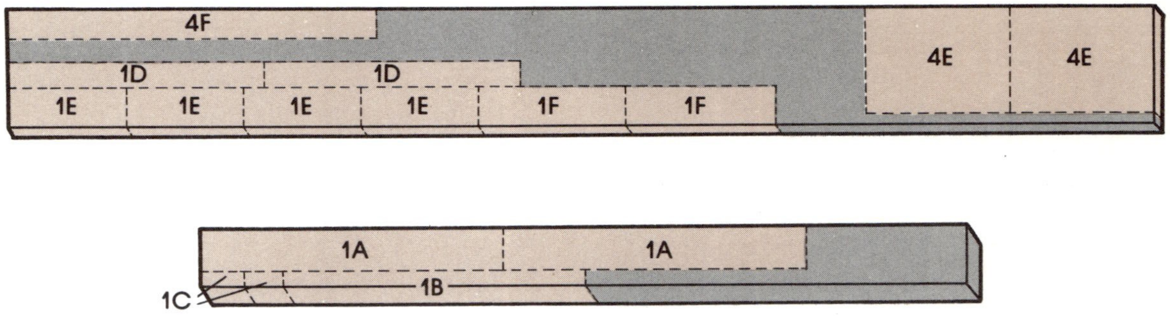

Illustration 5-1—The cutting diagram shows the labels and dimensions for pieces of the table-mount press to be cut from a mahogany 1 × 8 and a redwood 2 × 4. Scrap is shaded gray.

saw or a handsaw with a rip blade, cut along this line. Next, work with this strip of wood. Mark a line 7¼ inches from the end of the strip, and cut the piece off. Repeat three more times, and label all four identical pieces 1E. These are the legs, which will be notched and contoured later. Continuing on the same strip of wood, mark a line $9^3/_{16}$ inches from the end, and cut the piece off. Repeat and label both pieces 1F. These are the leg braces, which will be beveled on one edge in the assembly phase. You will end up with a small piece of scrap.

Now rip another strip from the original board, but this time cut into the board only as far as necessary for the required pieces. Draw a line 1½ inches from the end of the board for a distance of at least 32½ inches, and cut off the strip. Mark a line across the strip 16 inches from the end, and cut on this line. Cut another piece 16 inches long, and label both pieces 1D.

For the next rip cut, draw a line 1⅞ inches from the factory edge of the board for a distance of 23 inches. Cut the strip off at this point. This gives you a piece 23 inches long for the lever, which will be shaped later. For now, label this piece 4F, and put it aside. This completes the mahogany cuts. Now go on to the redwood 2 × 4 for the rest of the pieces.

There is one rip cut on the redwood 2 × 4. It must be cut for a distance of 38½ inches on the wide face of the board. Draw a line ⅞ inch from one edge for that distance, and cut along the line. Cut two pieces from this strip, each 2½ inches long. Label these pieces 1C. The next cut is 19 inches from the end. Label this piece 1B. From the remainder of the 2 × 4, which now measures 2½ inches wide, cut two pieces, each 19 inches long. Label both pieces 1A. This completes the cutting. With all the pieces cut, you are ready to assemble the table-mount press.

ARMS AND DECK ASSEMBLY

Before you can assemble the arms and deck of the table-mount press, certain pieces have to be notched and shaped. The two arms must be notched on the bottom. To cut those notches, draw a series of lines, and cut along these lines with a handsaw or saber saw. Look at illustration 5-2 for a detailed view of the notching. At a point 1¼ inches from the

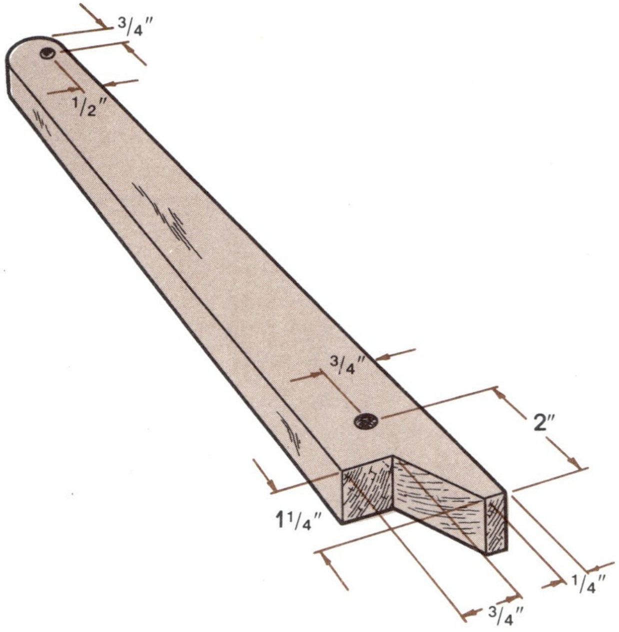

Illustration 5-2—The arms are shaped as shown. A dowel hole is drilled at the bottom end, and a clearance hole is drilled at the top end.

end of the arm, draw a line ¾ inch from the edge. Draw another line at an angle from the end of this line to a point ¼ inch from the opposite corner, as shown. Cut along these lines to take out the piece formed by the lines. It is a tricky notch, so be sure you have the lines drawn correctly before you begin to cut. Repeat this process with the other arm. This completes notching the bottom ends of the arms.

Next, drill a clearance hole for the bolt that will be inserted to bridge the two arms after the deck is assembled. Mark on one arm, then drill through both arms at the same time to make them identical. To locate the center point for the hole, mark a point ¾ inch from the top end of the arm and ½ inch in from the side, as shown in illustration 5-2. Put the two arms together, face to face, with the edges and the ends flush, and, using a ¾-inch spade bit, drill a counterbore hole on the center point, $^{1}/_{16}$ inch deep. Then, using a $^{5}/_{16}$-inch bit, drill a hole through both arms at the same time. Use the center point of the counterbored hole for the second hole. Be sure you only drill the $^{1}/_{16}$-inch-deep, ¾-inch counterbore on one piece. Then, with a countersink bit, countersink the hole on the opposite outside face of the arm that was not counterbored. Do this on the outer face where the $^{5}/_{16}$-inch bit broke through. Then round the top end of both arms to give a finished appearance.

Finally, mark center points on the other end of the arms. Take a good look at illustration 5-2 to make sure of the location of the center point, then measure 2 inches from the

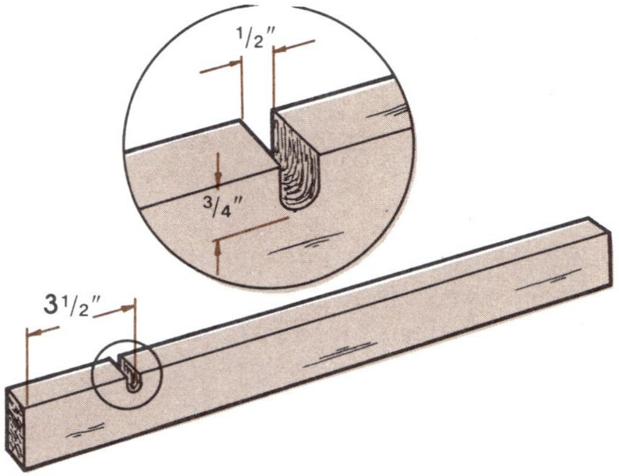

Illustration 5-3—The center rail is notched as shown, before the deck is assembled.

notched end on the face of the piece, and mark the center point ¾ inch from the edge, as shown. Drill a $^{5}/_{16}$-inch hole through the piece. Repeat this procedure for a center point on the other arm.

Put the arms aside for now, and find piece 1B, the center rail of the deck. As shown on blueprint sheet 1, it is notched to receive the copper tubing fitted on the top of the arms when they are in the folded position. Mark the center point for the hole 3½ inches from one end of the center rail and ½ inch from the edge, as shown in illustration 5-3. Using a ½-inch bit, drill a hole at the center point. Then make two small cuts from the edge to the sides of the hole to open a notch the same width as the diameter of the hole, as shown.

Now you are ready to drill the holes in the deck pieces for the threaded rods. This operation is similar to doweling a butcher block together, but instead of dowels you will use threaded rods. The fit is looser than dowels would be, making it easier for you to get the holes in all of the deck pieces aligned with each other.

First, mark the center points for all of the clearance holes in the deck pieces, then drill them. All of these pieces are drilled through their 1½-inch dimension. Mark the center points for the holes on the edge of the side rails and the face of the other pieces, as shown in illustration 5-4. When the pieces are put together, they will form a 1½-inch-thick deck. Before you begin, look closely at illustration 5-4 and blueprint sheet 1 to better understand the procedure.

Start with the side rails, 1A. Mark a center point ¾ inch from one end and another

Illustration 5-4—The deck of the table-mount press is assembled with clearance holes for the threaded rods aligned as shown.

center point 2¼ inches from the opposite end. Both center points are marked on the edge of the piece, ¾ inch from the face. Repeat this procedure on the other side rail. Next, work with spacers, 1C. On each of them, mark a center point on the face of the piece, ¾ inch from the end and ¾ inch from the edge. Now work with the center rail, 1B, and mark a center point ¾ inch from the end of that piece that is closest to the notch. Check illustration 5-4 to be sure you have the right end. The second center point on this piece is marked 2¼ inches from the opposite end. Both center points are on the face of the piece, ¾ inch from the edge.

Before you drill the holes, check again to be sure that your marks are correct. You will drill counterbore holes for the T-nut fasteners in the edge of one side rail and counterbore holes for the nut and washer combinations in the edge of the other side rail. Use a ¾-inch spade bit to drill ¼-inch-deep holes on the center points marked on one side rail for the washer-nut combinations, and counterbore only $^{1}/_{16}$ inch deep on the center points on the other side rail. Note that these holes do not get drilled all the way through the piece. They are merely counterbore holes to allow the T-nut fasteners and the washers and nuts to be sunk beneath the surface of the side rail. After these are drilled, change bits to drill the clearance holes through the pieces. Using a $^{5}/_{16}$-inch bit, drill through the rest of the pieces on each center point.

With all of the clearance holes drilled, take one of the side rails and tap a ¼-20 T-nut fastener into each $^{1}/_{16}$-inch-deep counterbore hole on the edge of the piece. The shank of the T-nut fastener will fit into the clearance hole. The barbs on the flange will go into the wood to keep the T-nut fastener from turning in the hole. Now cut two 7½-inch-long sections of ¼-20 threaded rod. Use a hacksaw to cut them, and be careful not to ruin the threads near the ends. The best way to do this is to mark the cut, and then turn a nut to each side of the mark, leaving just enough space for the saw blade to fit between the nuts. Make the cut, then turn the nuts off the ends that you've cut. Any irregularities in the thread caused by cutting will be corrected as you turn the nuts over those threads.

Now it is time to put the pieces together to form the deck of the table-mount press. Look at illustration 5-4 and blueprint sheet 1 again before you begin. Some pieces will be glued in place, so it is best to test-fit all the deck pieces together first. Insert one of the sections of threaded rod into the side rail with T-nut fasteners installed, and screw it into the T-nut fastener until it is flush with the outside face of the T-nut fastener. Insert the other threaded rod into the hole at the other end of the side rail, and screw it into the T-nut fastener in the same manner. The rest of the deck pieces will slide onto the rods, leaving just enough of the rods protruding into the counterbore hole in the other side rail for a washer and nut to be turned onto them. Glue will be applied to both sides of each spacer before they are finally positioned on the rod. On the other end of the deck, the arms will be flanked by washers. The washers space out this end of the deck so it is

Photo 5-1—A nut and washer are turned onto the threaded rod, fitting into the counterbore hole in the side rail. The nut is tightened with a ratchet wrench.

equal to the end with the spacers.

Line up the pieces as shown on blueprint sheet 1 and illustration 5-4. Make sure you have them in the right sequence. Slip a ¼-inch flat washer onto the threaded rod that is 2¼ inches from the end of the side rail, and fit the arm that is countersunk, NOT counterbored, onto the same rod. The countersink should be against the rail. Slip another ¼-inch flat washer onto the rod so that there are washers against both faces of the arm. Install the spacer on the other end. The spacers are the only pieces of the deck that must be glued to keep them from rotating on the threaded rod. Use any water-resistant wood glue. Read the directions on the container before you apply the glue. Apply glue to both faces of a spacer and slip it onto the threaded rod.

Then slip the center rail onto the rods against the spacer and the washer on the arm. The notch in the center rail will be adjacent to the hole in the top end of the arm. Slip a ¼-inch flat washer onto the threaded rod in the end with the arms. The second arm is then put onto the rod with the counterbore facing toward the exposed ends of the rods, and a ¼-inch flat washer is slipped onto the rod so that this arm is also flanked by washers. The second spacer is glued into place on the same rod as the first spacer. Apply glue to both faces of this spacer as well, and fit it onto the threaded rod. The final piece is the remaining side rail. Slide it onto the threaded rod with the counterbore holes facing out. Fit each threaded rod with a ¼-inch flat washer and a ¼-20 nut, as shown in photo 5-1. They will fit into the ¾-inch counterbore hole and can be tightened with a ratchet wrench, using a $^{7}/_{16}$-inch socket. All of the pieces of the deck should form a flat surface. Make sure the spacers are even with the other pieces and that both arms and the center rail are in properly before tightening the nuts. Let the deck sit while the glue dries.

The tops of the arms are connected by a bolt and a small section of copper tubing which forms a catch for the lever when applying pressure. Pull one arm into the upright position and hold it against a hard surface while you tap a T-nut fastener into the counterbored hole in the outer face of the arm near the top. Now, cut a $^{15}/_{16}$-inch-long section of ⅜-inch (outside diameter) copper tubing. Insert a ¼-20 flathead bolt 2½ inches long through the arm with the countersink hole, fit it with a ¼-inch flat washer, then slip the

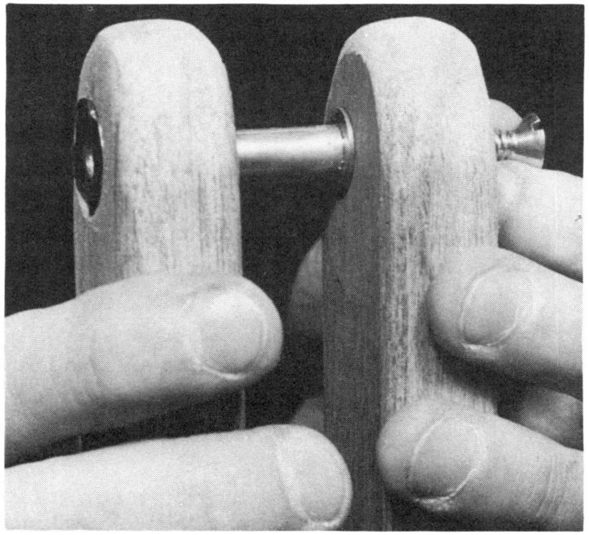

Photo 5-2—A flathead bolt is fitted into a hole in one arm and through a washer, a small section of copper tubing, and another washer, then screwed into a T-nut fastener on the outside face of the other arm.

copper tubing over the bolt, put on another ¼-inch flat washer, and push the bolt into the other arm, as shown in photo 5-2. Screw it into the T-nut fastener until it is tight. Do not tighten the bolt too much, or the copper tubing may buckle. The copper tubing will now form a bridge between the two arms, and it will fit into the notch in the center rail when the arms are folded down. This completes the deck and arm assembly.

LEG ASSEMBLY

There are two leg assemblies, each with two legs and a leg brace. Look at blueprint sheet 1 for details on these pieces and their positions on a completed press. They are identical, and both will be built in the same way. The legs are parallelograms with rounded corners, and the upper edge of the leg brace is beveled to fit flat against the bottom of the deck when the legs are unfolded. The leg brace fits into the edge of the leg by means of a notch, which is really a single dovetail joint. Don't let the term intimidate you. The outline for the joint is traced and can be easily cut out with an ordinary saw and a wood chisel. Take a look at illustration 5-5 for a view of a leg assembly.

We've provided a cutting pattern, illustration 5-6, for shaping the legs in order to simplify the cutting for you. You will be working with the four pieces labeled 1E. The cutting pattern should be copied or traced, so you don't have to tear the page out of the book in order to use it. Place the pattern copy on top of one piece labeled 1E, and trace its

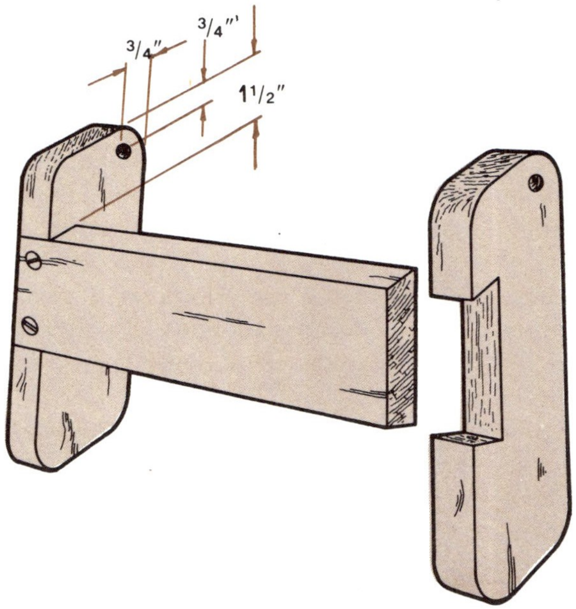

Illustration 5-5—The leg assembly, consisting of two legs and one leg brace, is assembled as shown. Note the position of the beveled edge of the leg brace and the position of the clearance holes for the pivot screw.

outline onto the face of the piece. The sides of the pattern should align with the edges of the wood, but the ends will have to be angled and rounded at the corners. Do not trace the notch for the leg brace yet. It is shown on the pattern merely to give an idea of its position. You'll get a better-fitting joint if you trace the end of the actual leg brace onto the leg. With the pattern drawn on each of the four legs, cut them out.

Drill clearance holes for the shank of the pivot screws in each leg. Mark a center point for the hole ¾ inch from the high point of the angled end and ¾ inch from the edge of the piece, as shown in illustration 5-5. Using a $^7/_{32}$-inch bit, drill through the leg at this point. Repeat this process on the other legs.

The best way to cut the bevel on the edge of the leg braces is to plane them at an angle. The angle is not very sharp, and only a little bit of wood needs to be taken off. First, mark guidelines for planing the bevel. On the face of the leg brace, draw a straight line ¼ inch from the edge for the length of the piece. Extend this line around the ends and angle it to the opposite corner of the edge, as shown in illustration 5-7. To do the planing, clamp the leg brace against the edge of a workbench or sawhorse. Hold the plane firmly and even with the angle of the bevel, checking all the lines as you plane and removing only a little wood at a time. Repeat this process for planing the bevel on the other brace.

With the legs and the leg braces prepared, the next step is to put them together. As stated before, this is a single dovetail joint, as shown in illustration 5-5 and on blueprint sheet 1. Study the blueprint sheet carefully to understand the joint and how it fits into the entire assembly. With the bevel planed on the leg brace, you already have the shape for the joint.

To begin, draw a straight line across the face of the leg 1½ inches from the end and parallel with the angled end, as shown in illustration 5-5. Now place the end of the leg brace against the face of the leg, with the beveled edge of the leg brace on the line and the narrower of the two faces of the brace flush with the edge of the leg, as shown in photo 5-3. Hold the leg brace in position, and, using a very sharp pencil, trace its outline onto the face of the leg, as shown. The line you've drawn will be concurrent with one line of your tracing.

Cut out the shape of the leg brace traced on the leg. A saber saw or coping saw is good for this job. When you cut the joint, be careful not to take out too much wood. Cut on the inside of the lines to assure a tight fit. Adapt the shape of the notch to the end of the leg brace with a wood chisel if necessary. Do one joint at a time, cutting it out and fitting the leg brace into it.

With a leg brace inserted into the notch so that the end of the brace is flush with the outer face of the leg, drill two pilot holes through the leg brace and into the leg, using a 5/64-inch bit. Then drill clearance holes for the screw shanks, ONLY INTO THE LEG BRACE and no deeper. Countersink the holes with a countersink bit, and fasten the joint with 1½-inch #8 flathead wood screws and glue. Do each joint in the same manner. If you complete one joint at a time, you can custom-fit each one separately, since one end of a leg brace may be shaped slightly different from the other end.

The next step is to fasten the leg assemblies to the deck assembly. The legs will fit against the outside of the deck and fold neatly to the deck for easy storage. You've already drilled screw-shank clearance holes in the legs to allow them to pivot on the screws. Now drill pilot holes for the pivot screws in the

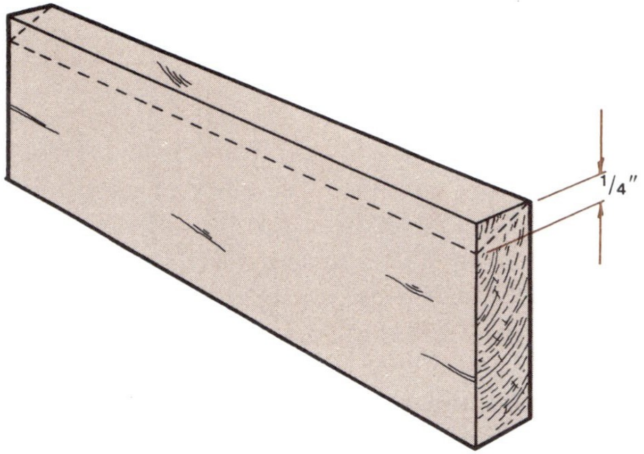

Illustration 5-7—Lines are drawn on the leg brace to mark the bevel. The wedge of wood will then be taken off with a plane.

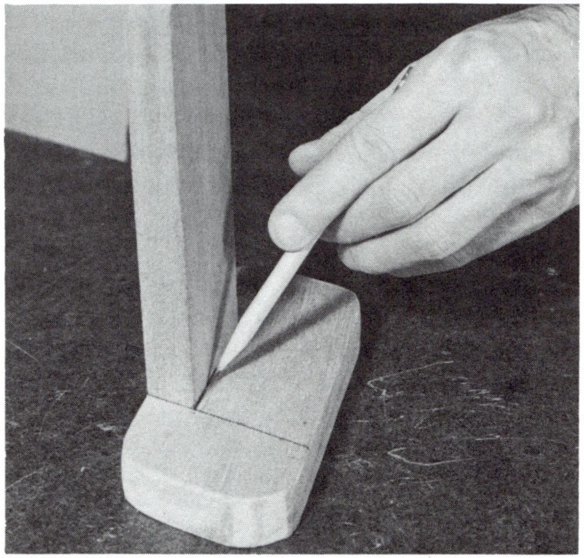

Photo 5-3—Trace the shape of the single dovetail joint onto the leg, using the end of the leg brace, keeping it flush with the edge of the leg and the bevel edge on the line parallel to the top of the leg.

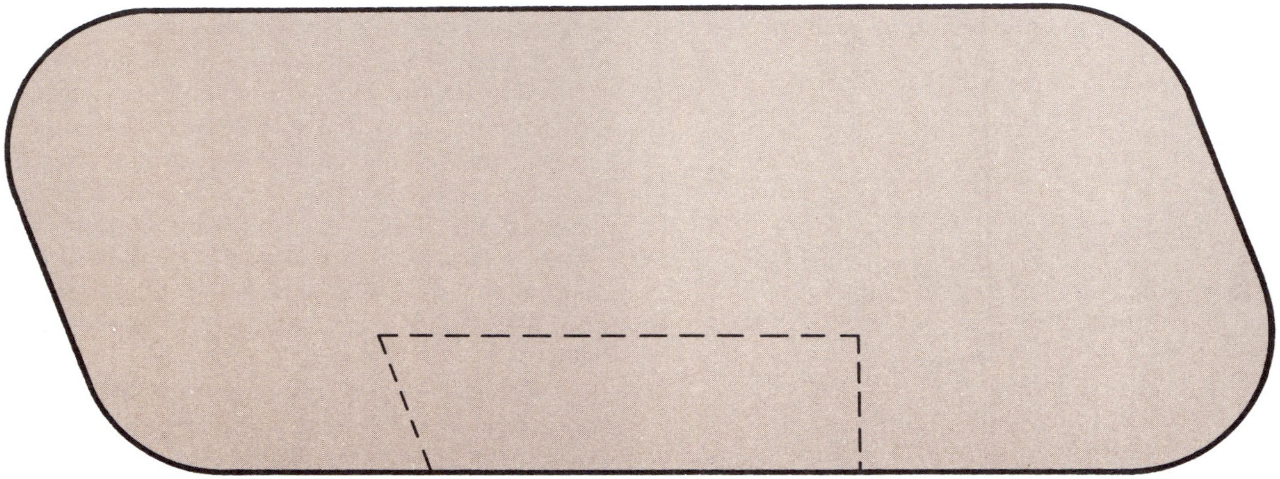

Illustration 5-6—Use the leg pattern for cutting the shape of the legs. Do not use the pattern to cut the leg brace joint.

edge of the side rails of the deck. First, mark the center points for the pilot holes. These are marked on the outer edge of each side rail, ¾ inch down from the face of the deck. For the first hole, measure 4½ inches from the end of the side rail where the arms are fastened. Then mark a point 2¼ inches from the opposite end of the side rail. Mark center points on the other side rail in the same manner, and drill pilot holes on each point, using a 3/32-inch bit.

Fit a leg assembly onto each end of the deck. The top bevel on the leg brace will fit flat against the bottom face of the deck when the legs are in the open position. The legs will fit loosely against the side rails. Match the shank clearance holes in the legs with the pilot holes in the side rails. Fit a 2-inch #12 roundhead wood screw with a ¼-inch flat brass washer, insert it into a leg, and screw it into the side rail. Do not tighten the screw all the way; just bring it to ⅛ inch from the face of the leg. Adjust this gap after installing a pivot screw in the other side of the leg assembly so that both legs are evenly spaced from the side rails. Install pivot screws in the other leg assembly in the same manner. The two leg assemblies should fold freely when you are finished. Do not overtighten the screws, as this could prevent them from folding properly.

LEVER AND PLUNGER

The lever for the table-mount press is almost exactly the same as the lever for the pot-mount press. The only difference is that the hole for hanging the plunger on the table-mount lever is in a different position. For the table-mount press, the hole for hanging the plunger is positioned 9 inches from the end of the lever, but still ½ inch from the lower edge, as shown in illustration 5-8. Other than this detail, the lever is the same. The plunger is exactly the same as the one for the pot-mount press. Follow the details given in chapter 4 in the sections "Shaping the Lever" and "Shaping the Plunger," beginning on page 54. When working on the lever, just change the position of the hole for hanging the plunger to 9 inches from the wide end of the lever.

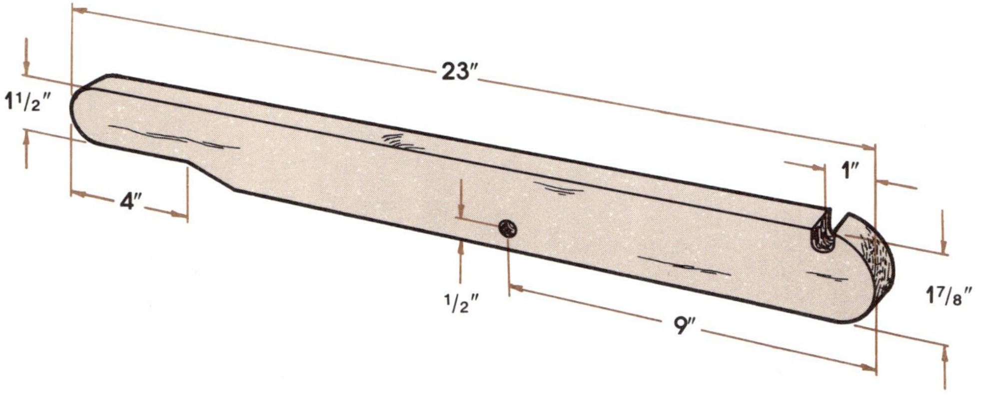

Illustration 5-8—The lever is contoured on one end to form a handle grip and rounded on the opposite end. The notch and hole position are shown also. Note that the hole for hanging the plunger is 9 inches from the notched end, as shown.

With the table-mount press, you'll need a shallow pan to catch the drippings from the tofu box. This pan should be at least 10 inches wide, 15 inches long, and about 1 inch deep. It sits on top of the press deck, sandwiched between the press and the tofu box. The soymilk or whey goes through the tofu box and ends up in the tray. The tray needs holes in it for the liquid to run through into a pot positioned below the tray. When you have chosen the tray that you will use with the table-mount press, drill four ¼-inch holes, ½ inch apart, in one corner of the tray. When drilling through metal like this, you will undoubtedly leave some sharp burrs around the holes. To take these off, use a countersink bit and drill at a slow speed, just grazing the edges of the holes.

When you are finished with the tray, your table-mount press is complete. Used in combination with the tofu box described in chapter 3, the table-mount press will help you with your tofu making for a long time to come.

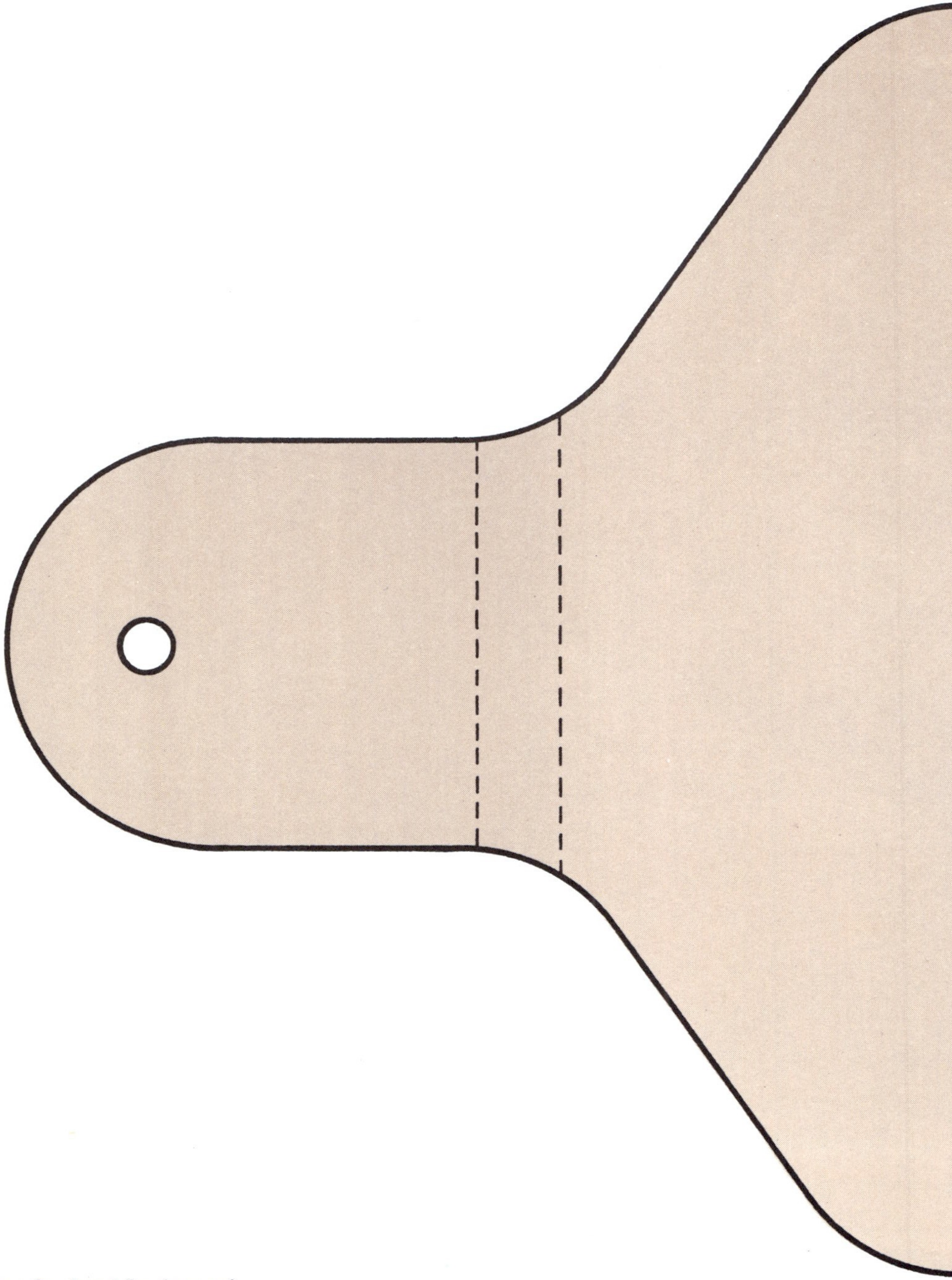

CUTTING THE PLUNGER

Cutting the curves of the plunger may look awesome at first, but this simple pattern can be traced on the two pieces labeled 4E that you've already cut for the plunger halves. You can make a carbon copy or a photocopy of the pattern so that you won't have to tear it out of the book, trace it on the plunger pieces, then just follow the pattern with a jig saw to cut out the shape of the plunger. When both halves are shaped, each one will be notched by a few simple cuts with a handsaw and glued together.

The grain on the plunger must run vertically, so before you begin tracing the pattern on each half of the plunger, make sure that the grain on the piece is running from the base of the plunger to the stem at the top of the plunger, as shown in illustration 4-8. The line for the base of the plunger can be concurrent with one 7¼-inch edge of the pieces to be cut. Then simply draw the rest of the lines from there on both pieces labeled 4E. Use a saber saw to cut out the shape of the plunger. If you don't have a saber saw, a coping saw will do the job. You should end up with two identical pieces, which you will notch and then glue together.

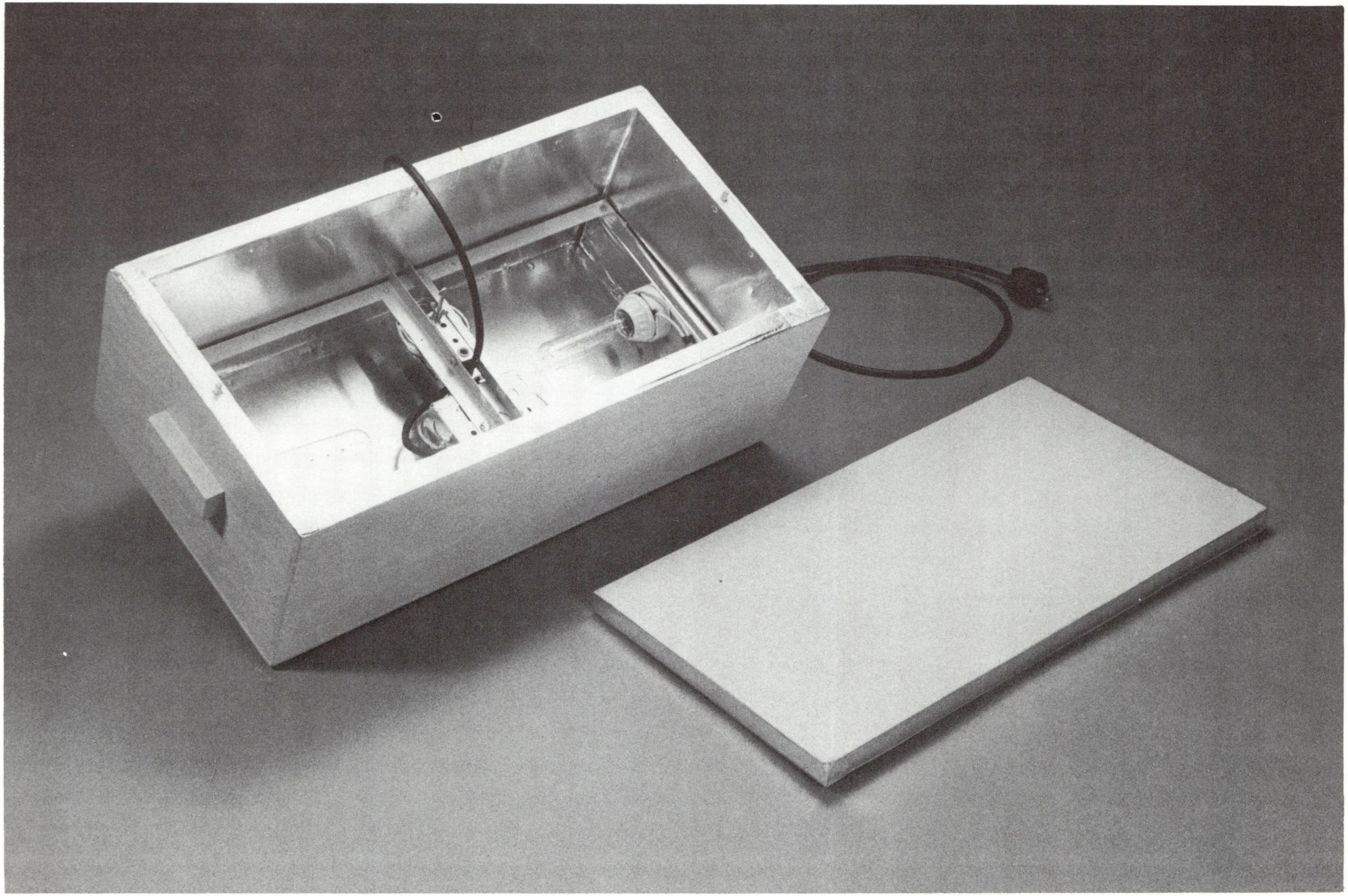

6 Tempeh Incubator

The tempeh incubator stands apart from the other projects in this book in that it is the only one which deals with tempeh.

What you will be building is a low-heat oven reaching and holding a contant temperature of around 92°F. The heat is supplied by two 25-watt light bulbs mounted to a frame on which the tempeh trays sit, inside the box. This entire frame with all of the electrical installations, including a thermostat to control the heat, can be lifted out of the box for easy, safe cleaning. The box itself is plywood on the outside with an inner framing and a layer of insulation. The entire inside of the box is covered with a reflective surface to help conserve heat.

Blueprint sheets 5 and 6 detail the incubator assembly. Take a good look at them before you begin.

Some of the materials for the tempeh incubator may not be easily identifiable. The box is insulated with Styrofoam insulation which is bought in sheets. Styrofoam is a brand name for a particular product. Don't confuse it with bead board or polystyrene types of insulation. It's light blue in color, and the name is stamped on the sheet. It is made by Dow Chemical, 2020 Dow Center, Midland, MI 48640 (phone 517-636-1000). Contact Dow Chemical if you have trouble obtaining the product. Chart 6-2 lists the amount of Styrofoam you need to insulate the incubator. A 2 × 8-foot sheet is big enough to give you all of the pieces you will need.

The materials list gives you the amount of aluminum flashing required for the whole box, and as you read through the chapter you'll find the sizes of each sheet that you need. Flashing comes in rolls of different widths, so just figure ahead how you will cut the needed pieces from the size roll you get.

Contact Paper is a second choice for the reflective material to cover the inside of the box. It is not as durable as the flashing, and it is not watertight, but it will serve the same purpose as the flashing. Contact Paper is a brand name for a particular type of shelf-lining paper. It is made by Comark Plastics, 3601 Hempstead Turnpike, Levittown, NY 11756 (phone 516-796-4100). The color is designated as Solid Chrome #2512 in the Contact Brand self-adhesive plastic catalog. It comes in 12½-yard rolls, but you'll need quite a bit less, so buy just the amount you will be using. Other companies may make a similar self-adhesive chrome paper, and this may be used also. The critical point is that the paper must have a chrome surface to reflect heat, and the surface should hold up after repeated wipings with a damp cloth.

The temperature regulator, or thermostat, used is made by Brower Manufacturing Company, 640 South 5th Street, Quincy, IL 62301 (phone 217-222-8561), and is available through the Sears, Roebuck and Company Farm and Ranch catalog. The catalog sales department in any Sears store will be able to help you order the thermostat, which is designed for a poultry incubator. It is listed in the 1981 Sears Farm and Ranch catalog as "Automatic Electric Temperature Regulator" and the order number listed is 32 AF 88022. The "AF" part of the catalog listing is for catalog identification, and changes periodically as the catalog is updated. If you do not have a Sears store near you, contact the Brower Company for direct mail order. The thermostat weighs 7 ounces and operates with 110–120 volt, 60-cycle AC, and 120 watts. The most important thing is that the temperature range for the thermostat is 75° to 125°F. If you decide to use a thermostat other than the one we recommend here, make sure that it can be used to keep the incubator at a

Tools

- Circular Saw, Saber Saw, or Handsaw
- Hacksaw
- Drill (with assorted bits)
- Hammer
- Screwdriver
- Utility Knife
- Plane
- Tape Measure
- Combination Square
- Awl
- Sandpaper
- Caulking Gun
- Assorted Wrenches

Chart 6-1—These are the tools you will need to build the tempeh incubator.

constant temperature of 92°F for making tempeh. You may have to change the position of the center brackets to accommodate the mounting of a different thermostat. Just be sure that it is located on the tray frame midway between the two light bulbs.

The box is built out of ¼-inch plywood. We have found luan plywood to be best. This is a solid-core plywood with a finished veneer on one side. It will paint better and will stain well, if you choose to finish the box that way. You can use regular ¼-inch plywood, but we feel the luan gives you a stronger box and a more attractive finished unit. Before buying your lumber, look at chart 3-3 on page 41, and be sure that you are getting the right actual size when you ask for nominal-size lumber.

CUTTING

The most important part of building the incubator is cutting the plywood pieces. Be extra careful to cut these square and to the correct size. These pieces are the shell of the box, and the frame members are fastened to them only as braces to drive your nails into to strengthen the box. If your plywood pieces are inaccurately cut, the whole box will be out of square.

Cutting the pieces is fairly simple. There are no fancy cuts to be made. You'll be cutting from ¼-inch plywood first, then two 1 x 2s and a 1 x 3 (all #2 pine), and a few

Materials

Description	Size	Amount
Lumber		
Plywood (good one side)	1/4″	4 × 4 feet
#2 Pine	1× 3	8 feet
#2 Pine	1× 2	16 feet (2 8-foot pieces)
Hardwood Dowel	1/4″ diam.	3″
Aluminum		
Right-Angle Molding	1/16″ × 3/4″ × 3/4″	6 feet
Flashing (Chrome Contact Paper may be substituted)	. . .	10 square feet
Aluminum Tape	2″	29 feet
Insulation		
Styrofoam	3/4″	2 × 8 feet
Hardware		
#8 Flathead Brass Wood Screws	3/4″	12
Common Nails	4d	48
16-Gauge Nails	3/4″	92
18-Gauge Nails	3/4″	78
10-24 Roundhead Stove Bolt	1 1/4″	2
6-32 Roundhead Stove Bolt	1/2″	4
1/4-20 Machine Screw	3/8″	2
10-24 Hex-Head Nut	. . .	2
6-32 Hex-Head Nut	. . .	4
1/4-20 Hex-Head Nut	. . .	2

Chart 6-2—These are the materials you will need to build the tempeh incubator.

Materials [continued]

Description	Size	Amount
Hardware [continued]		
Cake-Cooling Racks	10″ × 10¼″	2
Weather Stripping (self-sticking, closed-cell foam)	¼″ × ⅜″	73 inches
Rubber Feet or Bumpers (self-sticking or screw-mounted)	. . .	4
Panel Adhesive	. . .	1 tube
Electrical Supplies		
Light Bulbs	T-10, 25-watt	2
Porcelain Sockets	2-wire	2
Wire	#18	120 inches (4 30-inch pieces)
18-3 Electrical Cord (with ground)	. . .	6 feet
3-Prong Male Safety Plug	. . .	1
Wire Nuts (blue)	. . .	3
Insulated Staples	#10	4
Metal Junction Box (with cover)	4″ × 4″ × 1½″	1
Romex Connectors	To fit junction box knockouts	2
Thermostat (with 75° to 125° range—see page 67 for details)	. . .	1

pieces from ¾-inch by ¾-inch right-angle aluminum molding. Follow the cutting diagram in illustration 6-1 and refer to blueprint sheets 5 and 6, and you will have no trouble. Using a pencil, label each piece with a number and letter as you cut it. Mark and cut the pieces one at a time. If you draw all of your lines first before cutting, the pieces will not come out the right size because of the ⅛ inch, or so, of material removed by the thickness of the saw blade.

Make sure that the plywood sheet is square before you begin. Cut from a 4-foot by 4-foot sheet of ¼-inch plywood, good one side. This is plywood that has one side without knots, improving its appearance for finish work. The good side will face the outside of the box. You can also use ¼-inch mahogany plywood or any other veneered plywood if you wish. If you use a thickness other than ¼ inch, the measurments will change. Cut the pieces, using a saber saw, with the bad side of the plywood facing up. This way you will not have ragged and chipped edges on the good sides of the pieces after cutting.

To begin, draw a straight line, 8¾ inches from the edge, the length of the sheet, running with the grain of the veneer. Cut along this line. From this section, cut one piece 24⅛ inches long and label this piece 5C, and another piece 12½ inches long and label it 5B. Repeat this process to get two more 8¾-inch wide pieces, one 24⅛ inches long and the other 12½ inches long. Label them the same as the first two.

Now measure 24⅛ inches from one end of the remaining section of plywood, draw a straight line across the grain at this point, and cut along the line. From this 24⅛-inch-long section, cut with the grain to get two pieces, each 13 inches wide. Label them both 5A. This completes the plywood cutting.

You will cut four identical pieces from the first 8-foot 1 x 2. Mark a straight line 22⅛ inches from one end of the 1 x 2, and cut the piece off at that point. Cut three more pieces the same length, one at a time, and label all four 5F. Now go to the second 8-foot 1 x 2 and cut four pieces, each 9½ inches long. Mark and cut them one at a time. Label them all 5E. Next, cut four pieces, each 8¾ inches

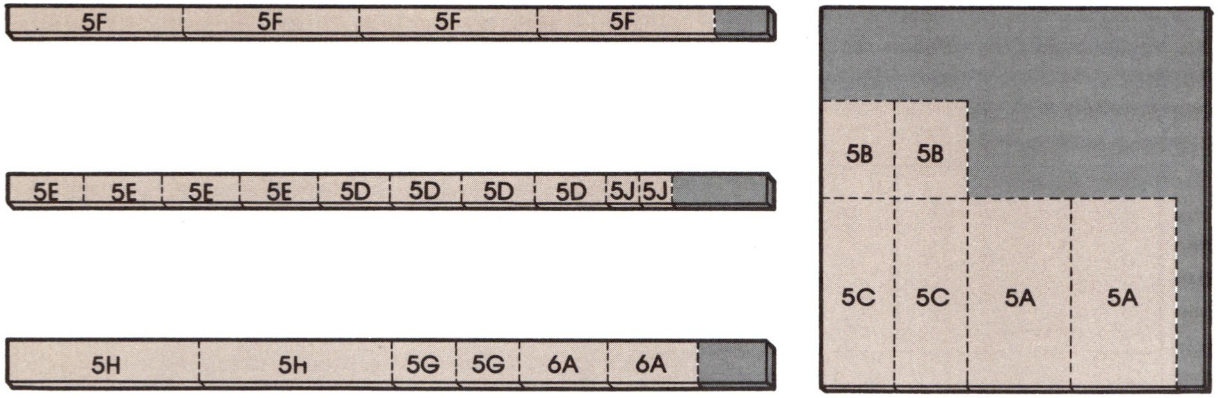

Illustration 6-1—The cutting diagrams show dimensions and labels of pieces for the tempeh incubator. The pieces are cut from two 1 × 2s, a 1 × 3, and a 4-foot by 4-foot sheet of ¼-inch plywood.

long, and label all four 5D. From the remaining section of 1 x 2 cut two identical pieces, each 4 inches long, and label these 5J. These two pieces will be beveled on one edge with a plane later to form handles for the box.

The last piece of wood to cut is a 1 x 3. From this piece cut two pieces, each 24⅛ inches long, and label them both 5H. Cut two 8-inch-long pieces, and label them 5G. Finally, cut two pieces measuring 10½ inches. Label them 6A.

You'll need a hacksaw for the next cuts, because you'll be cutting aluminum. From a 6-foot length of $^1/_{16}$-inch-thick right-angle aluminum molding, with each side measuring ¾ inch wide, cut two identical pieces, each 21⅝ inches long. Label them both 6B. Cut two more pieces 10½ inches long, and label both pieces 6C. This completes the cutting sequence for the incubator. To get a good fit, you will cut the insulation and the reflective covering for the inside of the box later as you assemble the box.

SIDE AND END PANELS

The sides and ends of the box are put together first by attaching a frame of 1 × 2 material to the plywood pieces to form panels, then fitting insulation into the framework, and finally applying a reflective surface over the insulation and framework.

Start with the two plywood ends, 5B, working with one at a time. Find the two 1 × 2 frame members labeled 5D and the two labeled 5E. The uprights, 5D, are fastened first. Apply panel adhesive to the face of one upright, then position it on the bad side of the plywood, as shown in illustration 6-2, flush with the 8¾-inch edge. The ends of the upright must be flush with the top and bottom edges of the end panel, as shown. Holding the upright in place, turn the panel over with the good side facing up, and, making sure that the upright is still in place, nail the panel to the upright, as shown in photo 6-1, using ¾-inch 18-gauge brads spaced evenly. Fasten the panel to the other upright, 5D, in the same manner, keeping the upright flush with the opposite edge of the plywood. The short parallels, 5E, fit between the uprights on the end panel, one flush with the bottom edge and one flush with the top edge, as shown in illustration 6-2. Fasten them in the same manner, first applying panel adhesive, then

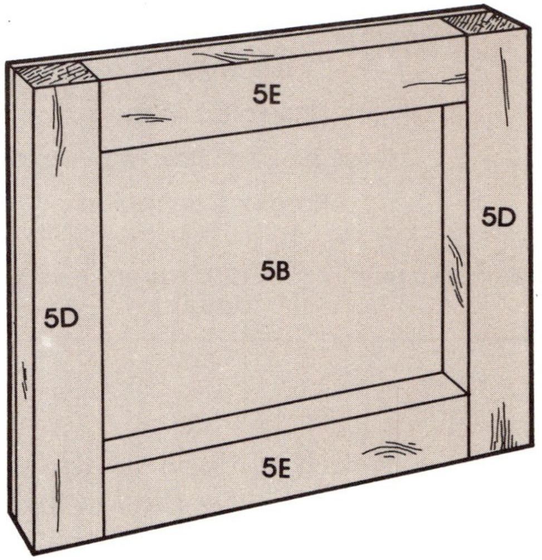

Illustration 6-2—Uprights, 5D, and short parallels, 5E, are positioned on the end, 5B, as shown.

Illustration 6-3—The handles are beveled as shown. The shaded area is taken off with the use of a plane.

Photo 6-1—Nail the upright in place on the end piece from the opposite face of the plywood. All of the frame members for the box and the lid are fastened in the same manner.

nailing from the opposite side with ¾-inch brads at even intervals. Repeat this process to fasten uprights and short parallels to the other end panel.

At this point, you should fasten the handles to the end panels. Find the two short pieces of 1 × 2 labeled 5J. To become handles, these blocks need a slight bevel on one edge to provide a better hand grip. To do this, mark a point ⅜ inch down from one edge on the end, and connect the point with the corner, as shown in illustration 6-3. Then plane off the material outside the line.

The handles are attached with panel adhesive and wood screws from the inside of the end panel, so that the screws will not show. Center each handle from side to side on an end panel, with the top of the handle 3 inches from the top edge. The beveled area goes toward the bottom of the panel, as shown on blueprint sheet 5. Trace the handle position on the wood, and drill two holes through the plywood in the handle area with a $^{5}/_{32}$-inch bit. Then reposition the handles, and, using the holes in the plywood, mark center points on the handles. Drill pilot holes into the handles with a $^{1}/_{16}$-inch bit. Be careful not to go all the way through the handle with the pilot holes. Sand and smooth the handles, and then attach each with panel adhesive and two ¾-inch #8 flathead wood screws.

Now go to the two side pieces, 5C. You

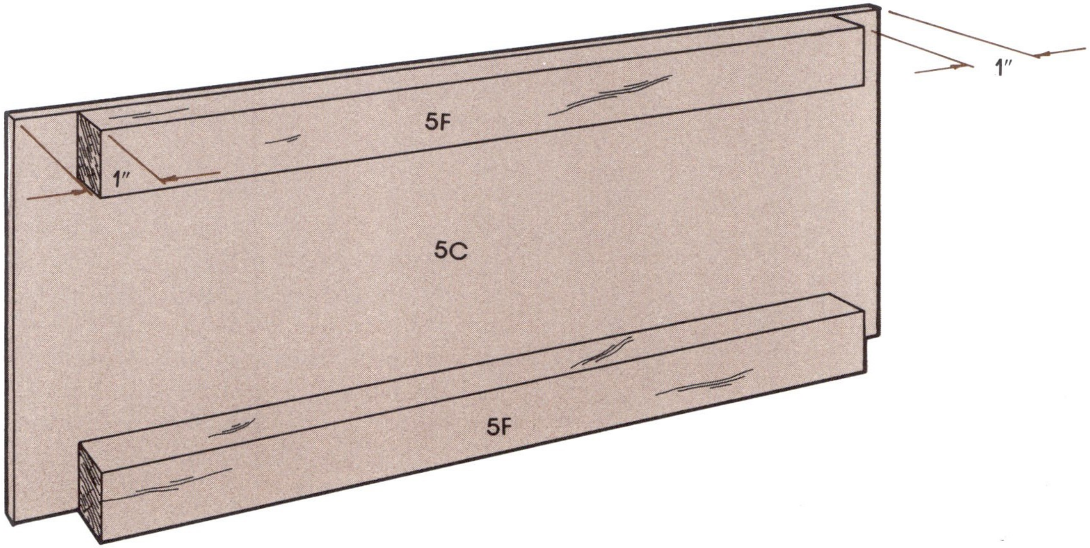

Illustration 6-4—The long parallels, 5F, are positioned on the sides, 5C, flush with the top and bottom edges and 1 inch from the ends of the side piece.

will install long parallels, 5F, on each side in the same way that you installed the short parallels on the ends. The side panels do not need uprights. To begin, draw a line 1 inch from each end of the plywood piece. Position a long parallel between the lines, flush with the bottom edge, and another flush with the top edge, as shown in illustration 6-4. Fasten them in the same manner, with panel adhesive and ¾-inch 18-gauge brads spaced evenly. Repeat for the second panel.

Photo 6-2—Position the insulation in the space between the frame members. On the side panels, keep it flush with the ends of the long parallels, as shown.

Now you will fit the end panels and side panels with insulation. The materials list gives the total amount of insulation that you will need for the project.

Use insulation that is ¾ inch thick, the same thickness as the frame members. If you cut it to the exact size, the insulation will neatly fill the space within the framing of the panels. Cut the insulation, using a sharp utility knife, so that you have a tight fit. Although the sizes for the panels will be given, it is best to check these sizes with the actual panels you've built, to allow for any slight differences. Each sheet of side-panel insulation will measure approximately 5¾ inches by 22⅛ inches, and each sheet of end-panel insulation will measure 5¾ inches by 9½ inches. Spread panel adhesive on each sheet of insulation and position it in the panel, as shown in photo 6-2. Make sure that the edges of the insulation are pushed all the way in so that the insulation is flush with the frame members. On the side panels, the insulation will not go past the ends of the long parallels, as shown in photo 6-2.

There are two types of reflective coating that you can use for the interior faces of the panels. Aluminum flashing is recommended because of its durability and because it is waterproof, but chrome Contact Paper may also be used. The flashing is installed as shown in photo 6-3. To cut the aluminum flashing, score it well with a utility knife, and bend it back and forth along the score line until it separates. Cut pieces to fit each panel, covering the insulation and the faces of the frame members. Cut two pieces of flashing 8⅝ inches by 12⅜ inches for the end panels, and two pieces 8⅝ inches by 22 inches for the side panels. You won't have to fit the aluminum flashing exactly, because the corners and edges of the box will be covered with aluminum tape later. Just don't make it larger than the framework on the panel. Use ¾-inch 16-gauge nails to nail the flashing pieces to the frame members. An easy way to start the nails is to poke a small pilot hole for each nail through the aluminum flashing before you nail. This can be done with an awl or a large nail. Space the nails evenly on each frame member. Note that the flashing does not go past the ends of the frame members and the insulation on the side panels. If you decide to use chrome Contact Paper instead, you will not need nails, as the

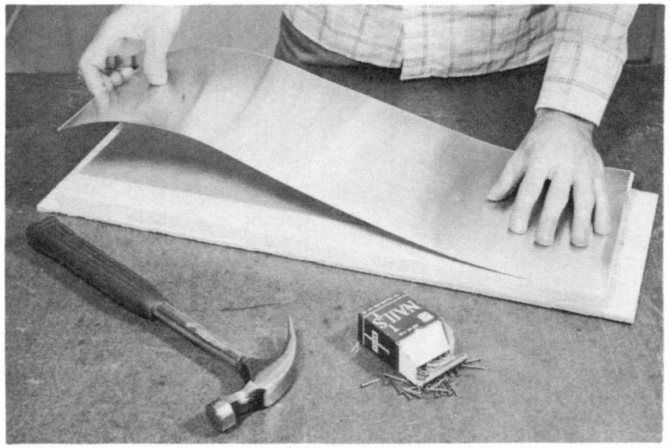

Photo 6-3—The flashing is installed on top of the frame members and the insulation. Do not let it overhang the frame members.

Contact Paper has an adhesive back that is self-sticking.

BOX ASSEMBLY

Now you have four completed panels: the two sides and the two ends of the incubator box. You'll see the incubator begin to take shape when you put them together and fit the bottom. The long parallels and insulation on the side panels are centered, leaving 1 inch of plywood on each end. This plywood overhang will overlap the edge of an end panel to form the corner joints of the box, as shown in illustration 6-5.

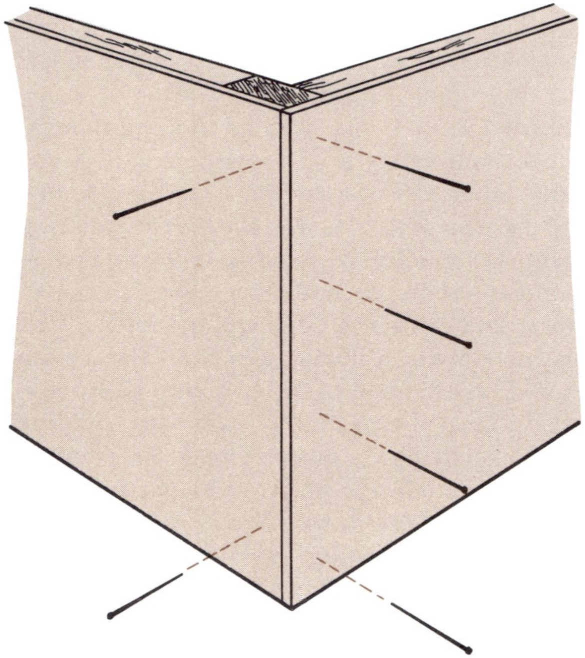

Illustration 6-5—The side panels overlap the end panels, as shown, to form the box.

Put the four panels together to make the rectangular box, applying panel adhesive and keeping the top and bottom flush. Drive 4d finishing nails through the plywood overhang of the side panel and into the upright on the end panel. Space four nails evenly along the joint. Then strengthen the joint by driving a 4d finishing nail through the end panel and into the ends of each long parallel, as shown. Check that the joints are square as you go. Be sure both handles face the same way. When the four panels are put together, check the square of the whole box with the bottom piece. The box must align exactly with the bottom.

Installing the bottom is tricky. Depending on what type of reflective surface you have chosen, you have to proceed in one of two ways. First we'll go over the instructions for Contact Paper, then we'll go over the special instructions for flashing. Test the assembled box on the bottom piece of plywood to make sure everything aligns.

Photo 6-4—Put the insulation in through the bottom of the box before the bottom is installed.

If you are using Contact Paper, simply cut a piece of insulation to fit exactly inside the box. The piece should be approximately 11 inches by 22⅛ inches, but cut to fit your exact opening.

With the bottom insulation cut to fit, cover one side of the insulation with Contact Paper and insert the bottom into the box, with the Contact Paper facing up. Push the insulation in through the bottom into the box before you put the bottom on, as shown in photo 6-4. Make sure you are working on the bottom of the box. The only way you can tell which is the bottom of the box is by the handles on the end panels. The bevels on the handles should face the bottom.

When the insulation is in the box, spread panel adhesive over the entire bottom face of the insulation and along the bottom edges of the box. Then fit the bottom in place, keeping it flush with the outer face of the sides and the ends. If everything is square, nail it in place, driving finishing nails through the bottom and into the frame members of the box. Then turn the box over and press the insulation against the bottom. Stack books on the bottom insulation until the adhesive sets up.

If you are using flashing, you must cut two additional frame members to provide a nailing surface for the flashing. These members are not included on the blueprints or the materials list. You don't need to nail the flashing on all four sides, only on the two long sides. Cut two additional pieces of 1 × 2 material, each 22⅛ inches long.

The extra frame members should be positioned as shown in illustration 6-6. They go 1 inch in from the ends and the long edges. Glue and nail the strips in place as you did for the other frame members. Test-fit the bottom with the frame members attached, to see if it will fit into the space between the four sides. If it fits, cut insulation to fit exactly between the frame members. If it does not fit, pull the pieces apart from the bottom, and adjust their position.

Apply panel adhesive to the insulation, and affix it to the bottom. Then, cut a piece of flashing 22 inches by 10⅞ inches, ⅛ inch smaller on each side than the outside dimensions of the frame members. The flashing should fit over the entire area of the frame members and the insulation. With the flashing cut, nail it to the frame members. Nail it with the spring curve of the metal as shown in illustration 6-6. This is important, because if you were to have the curve going the other way, it would have more of a tendency to pull up from the nails. That completes the bottom panel. Apply panel adhesive to the bottom edges of the four box sides, and glue and nail the bottom in place. You should now have a box with reflective material on all four sides and the bottom.

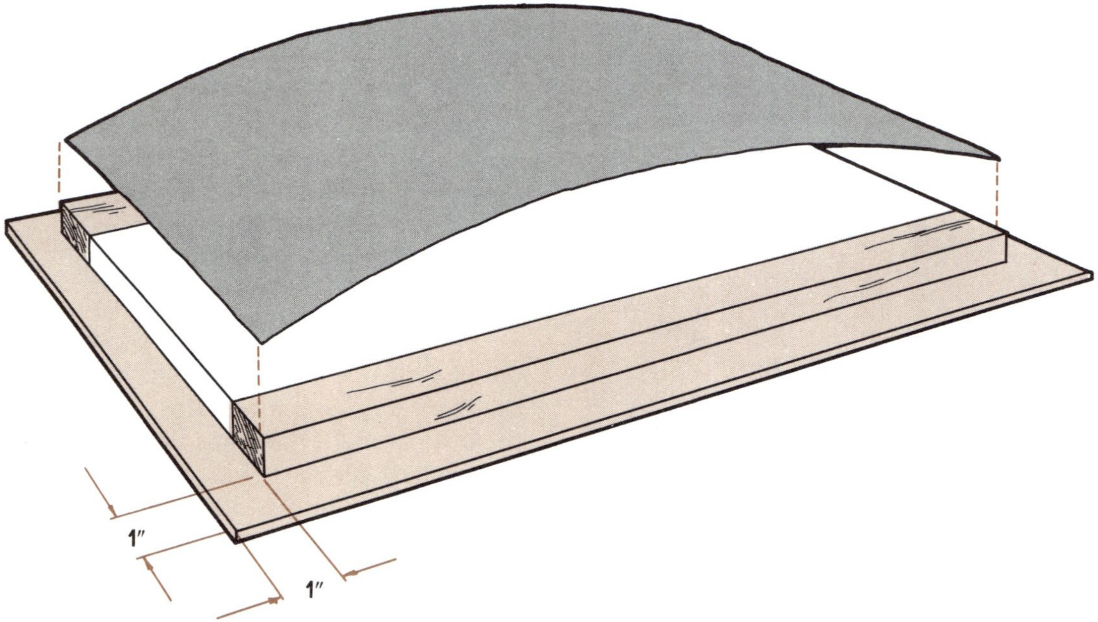

Illustration 6-6—If aluminum flashing is used, the bottom assembly will include extra frame members positioned 1 inch from the edges and the ends of the bottom. Insulation will fit between them and flashing on top, flush with the edges of the frame members. Note the curve of the flashing.

LID ASSEMBLY

The lid is free of the box, not fastened or hinged to it. You merely place it on top of the box and align the positioning pegs. It is assembled in the same way as the end panels, but without a handle. The remaining plywood piece labeled 5A is the lid. Two pieces, 5G, and two pieces, 5H, are the short and long casings respectively. Apply panel adhesive to the face of one piece of long casing and position it flush with the edge and the ends of the top, as shown in illustration 6-7. The casing should be attached to the bad side of the plywood. Turn the assembly over and, holding the long casing in place, check its position again and nail the top to the casing with ¾-inch 18-gauge brads. Space brads evenly along the top, driving them through the top and into the casing. Repeat this procedure for installing the other piece of long casing, 5H, to the opposite side of the top. The short casings should fit between the long casings, one on each end, flush with the end of the top, as shown in illustration 6-7. Fasten it the same way you fastened the long casing, using brads spaced evenly.

Now cut a sheet of ¾-inch-thick insulation to fit in the space between the casing members. It should be approximately 8 inches by 19⅛ inches. Check the dimensions of the space on your actual lid to get a tight fit. Fasten it to the inside of the lid panel with panel adhesive. Cut a sheet of aluminum flashing or chrome Contact Paper 12⅞ inches by 24 inches to fit the inside of the lid from end to end and side to side. Don't cut the flashing too big, or it will overhang the edges. File the sharp edges of the flashing, and nail it in place with ¾-inch 18-gauge nails spaced evenly along all the casing members.

TAPING AND DOWELING

All of the edges and inside corners of the box and the edges of the lid should be taped with aluminum tape to seal places where the pieces of flashing or Contact Paper join each other. Get 2-inch-wide aluminum tape, and seal all areas where two or more panels meet on the inside of the box. Illustration 6-8 shows a properly taped corner of the box. The best way to tape corners is to cut the tape to the right length, and fold it down the center with the sticky side out. Then push the tip of the fold into the corner, as shown in photo 6-5, and press it down along the joint. Press the tape down on both sides of the fold when you have it properly positioned.

Before you tape the top edges of the box, cut a U-shaped notch in the top edge of one side panel for the power cord. This can be done quickly and easily with a hacksaw. Center the notch from end to end, making it ⅜ inch wide at the top and ⅜ inch deep. Cut a section of tape about 1 inch square, and tape the inside of the notch. The tape you will

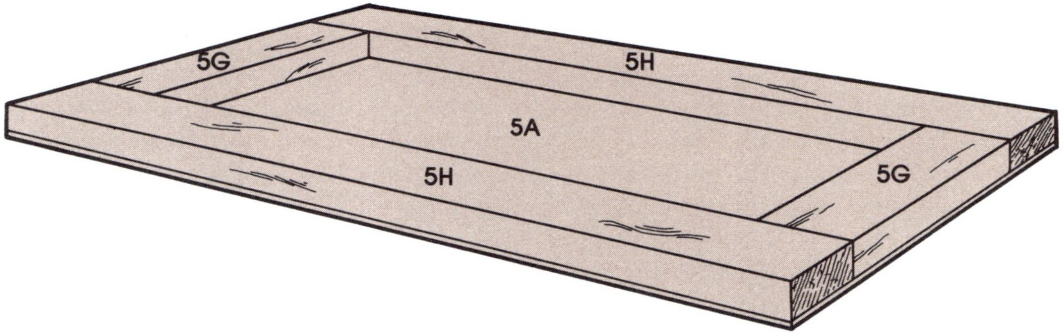

Illustration 6-7—The lid panel is formed by two long casings, 5H, and two short casings, 5G, fastened to the top, 5A, as shown. Note that the casings are flush with the edges and the ends of the top piece.

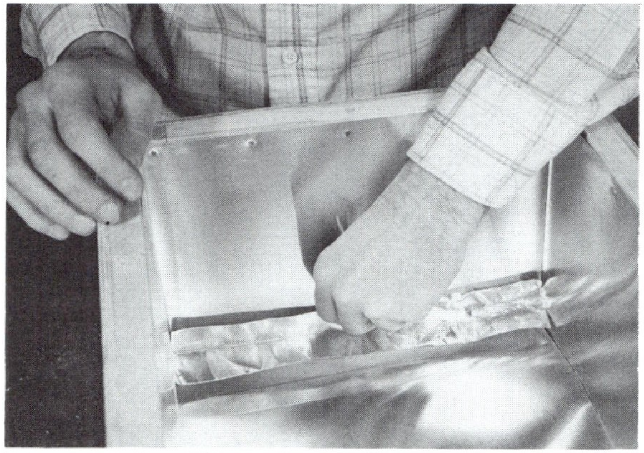

Photo 6-5—The aluminum tape is folded down the center, then placed in the corner, as shown.

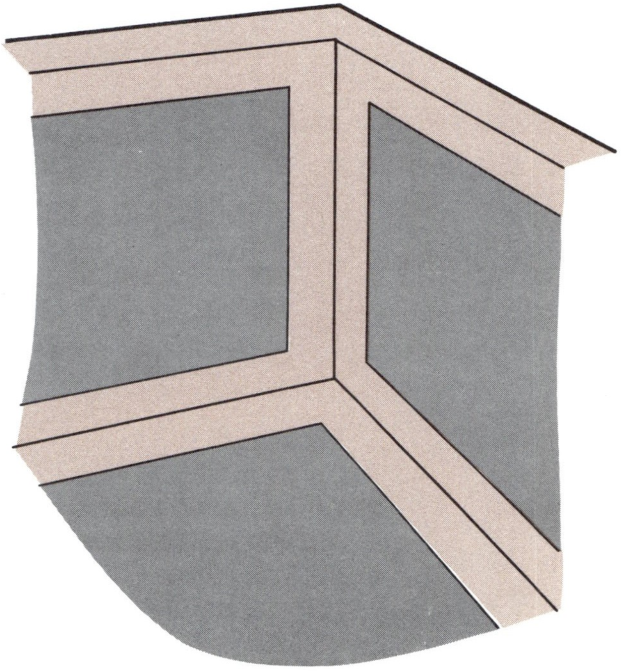

Illustration 6-8—The interior of the box is taped with aluminum tape, as shown.

put along the entire side edge will overlap this. Before you tape the top edges of the box, file the flashing edges so they are not sharp. Then tape the top edges on all sides.

Keep the tape flush with the outer face of the box, and lap the tape over onto the inside. Do one panel at a time. Cut the tape to the length of the outside dimensions of the box, so that the tape on the side panel will overlap the tape on the end panel at the corner. Where the tape crosses the notch, slit it in the center of the notch and fit the tape inside the notch. Now tape the edges of the lid, keeping the tape flush with the top face of the lid and folding it around the corner of the edge to cover the aluminum flashing or Contact Paper.

Now install dowels in the top edge of each end panel, and drill a hole into the underside of the lid in the same position, so that when the lid is placed on the box, the holes match the dowel pegs and hold the lid in place. These are the positioning pegs, 5K, mentioned earlier. Measure 6½ inches from one side of the box along the top edge of the end panel. Mark a center point for a hole at this point, ⅝ inch from the outside face of the end panel. Using a ¼-inch bit, drill a hole about 1 inch deep in the top of the end panel. Repeat this process to drill a dowel hole in the opposite end panel. Then on the underside of the lid, mark a center point for a hole 6½ inches from one side edge of the lid and ⅝ inch from the end. Drill this hole in the lid, using a $^{5}/_{16}$-inch bit. Drill this hole only ⅜ inch deep; do not drill all the way through. This will be a blind hole that won't be seen when the lid is in place. Repeat this process for the clearance hole in the other end of the lid, remembering not to drill all the way through the lid. When you have these holes drilled, go back to the box. Put a bit of panel adhesive into each hole in the box and tap a small section of ¼-inch-diameter hardwood dowel into each of the holes. After the dowels are firmly lodged to the depth of the hole, cut both off ⅜ inch from the edge of the end panel of the box. With a piece of fine sandpaper, round the tops of the dowels.

Apply weather stripping to the perimeter of the lid. Use ⅜-inch-wide, self-sticking, closed-cell weather stripping that is ¼ inch thick, and apply it like tape on the inside face of the lid along all four sides, keeping it flush with the edges. This will provide an added seal when the lid is in place, and it will also give the lid a nicer fit.

Finally, to complete the incubator box, fasten rubber feet, or bumpers as they are sometimes called, to the four corners of the bottom of the box. Use either self-sticking feet or the screw-mounted variety. These small rubber pieces, one on each corner, will provide a finished base for the box and keep it from marring your kitchen counter tops.

TRAY FRAME ASSEMBLY

Now that you have the incubator box assembled, you are ready to equip it with the working unit, which will provide heat for the incubator and a platform on which to place the trays of tempeh. This is the tray frame on which you will place cake-cooling racks to hold the tempeh pans. All of the electrical installations are made on the underside of the frame. It is a simple structure made of wood and aluminum molding, built as a separate unit so that it can be removed for cleaning the inside of the incubator box. Take a look at blueprint sheet 6 to get an idea of how the frame works.

You've already cut the pieces for the frame: two pieces of 1 × 3 labeled 6A, four pieces of ¾-inch by ¾-inch aluminum angle molding, two side rails labeled 6B, and two center brackets labeled 6C. This is all you will need to build the frame. First take the two 6A pieces, the end blocks, and wrap them in aluminum tape so that these pieces will reflect heat. Now you are ready to assemble the pieces. The side rails will bridge the end blocks, as shown in illustration 6-9. The right angle of the side rail fits around the corner of the end block and is fastened in place with a ¾-inch #8 flathead wood screw. Position the end of the side rail so it is flush with the face of the end block. Mark the center point for a screw hole in the side rail so that it is in the center of the edge of the end block and, using an $^{11}/_{64}$-inch bit, drill a clearance hole for the shank of the wood screw through the side rail. Use a countersink bit to countersink the hole. Put the side rail back in position, and mark a center point for the screw pilot hole in the edge of the end block. Use a $^{1}/_{16}$-inch bit to drill the pilot hole. Fasten the pieces with a ¾-inch #8 flathead wood

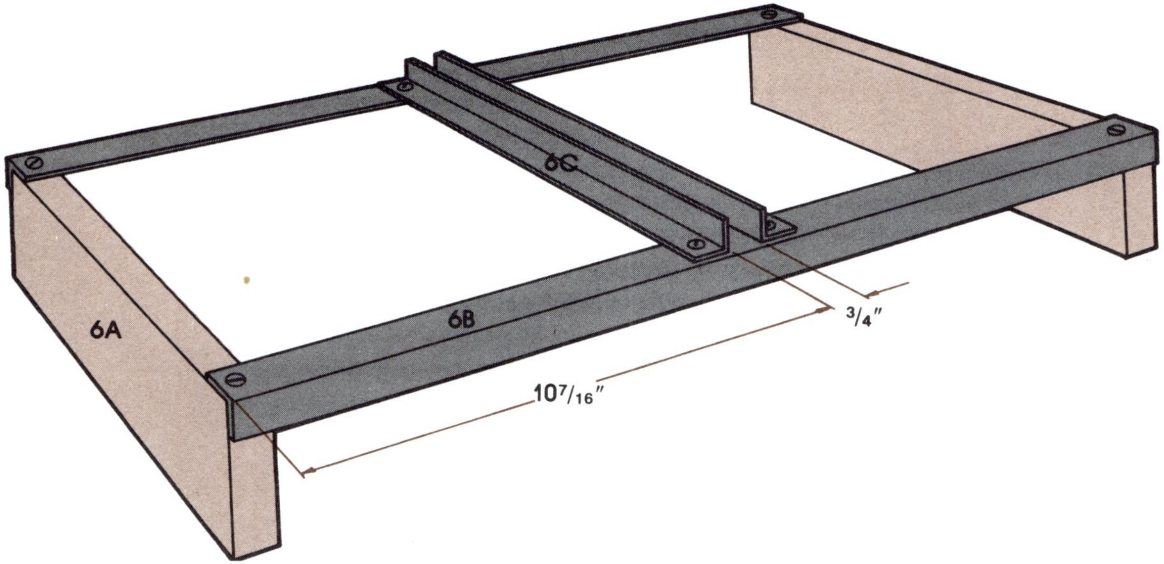

Illustration 6-9—The tray frame is made of end blocks, 6A, side rails, 6B, and center brackets, 6C. Note the dimensions for positioning the center brackets.

screw. Repeat this process with the opposite end of the side rail on the other end block. Then fasten the other side rail to the opposite corners of the end blocks in the same manner.

Two center brackets, 6C, also cut from aluminum angle, must be installed to complete the frame. They will be positioned across the side rails, as shown in illustration 6-9. Mark a point on the side rail $10^7/_{16}$ inches from one end. Mark another point ¾ inch from the mark you just made. The space between these two points should be just wide enough to mount your thermostat. Repeat this procedure on the other side rail. You will fasten a center bracket to both side rails with the corner of the angle on these marks. Hold the piece across the side rails with one side of the angle upright. Be sure that the corner of the angle is positioned on the marks. Holding it in place, mark a center point for a clearance hole on each center bracket. Using a $^5/_{32}$-inch bit, drill holes through the center brackets and the side rails. Now fasten the center bracket at each end to the side rails with a ½-inch-long 6-32 roundhead stove bolt in each hole, tightened with a 6-32 hex-head nut. The frame is now completed. The next step is to install the electrical parts of the unit.

ELECTRICAL INSTALLATIONS

There are four phases to the electrical assembly. First, you will attach a light socket to the inside face of each end block. Next, you will fasten a metal junction box to the center brackets to contain the wire connections. The third step is to fasten the thermostat to the center brackets, and finally, with all of these components installed, you will wire the system.

Before you do any of this, make sure you know what you are doing. Read through the following details beforehand, and study blueprint sheet 6. A view of the entire electrical installation is shown in photo 6-6. If you have no electrical experience, get some help. This is serious business, and a faulty hookup could be hazardous. A MISTAKE HERE COULD RESULT IN SERIOUS ELECTRICAL SHOCK.

The first thing you will do is to fasten the porcelain sockets, 6D, to the inside faces of the end blocks. These are standard two-wire porcelain sockets. Each socket will be centered end to end and top to bottom on the end block. Remove the cover on the socket, and position it on the end block with the terminals in a vertical position. Then mark the center points for screw pilot holes on the end block through the mounting holes in the socket. Using a $^1/_{16}$-inch bit, drill the pilot holes, and fasten each socket with two ¾-inch #8 flathead wood screws. Repeat this procedure to fasten a porcelain socket to the other end block. The sockets will face each

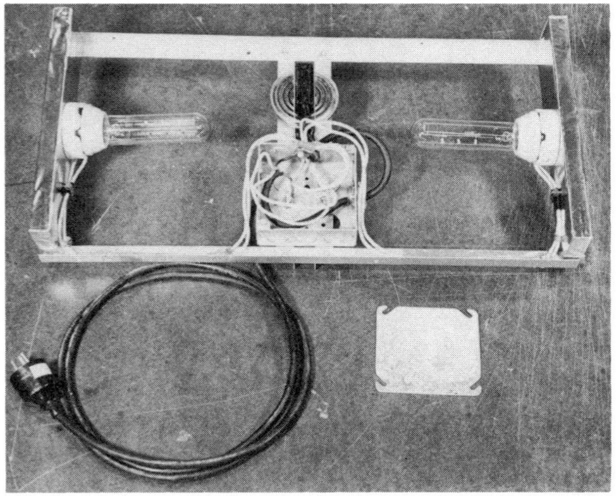

Photo 6-6—All electrical installations and wiring are done on the underside of the tray frame, with all wire-to-wire connections inside the junction box. The cover is then put on the junction box.

other so that the light bulbs will protrude toward the center of the frame. Use two 25-watt T-10 light bulbs with the unit. These are tubular-shape bulbs, as shown on blueprint sheet 6.

For safety, all of the wire connections for the unit must be put inside of a metal box, 6F, mounted on one side of the frame. This is the same type of box used by electricians for outlets and switches when wiring a house, but this one will have a plain, flat cover with no provision for switches or outlets. Use a 4-inch-square box, 1½ inches deep, and a cover to fit it. It will be mounted directly to the center brackets of the frame.

The junction box has several circles almost totally cut through the sides and back. These are called knockouts and can be easily pried out with a screwdriver or some other tool to open a hole in the box for incoming wires. Open two knockouts: one for the cord from the power source and the other for all of the wires to and from the thermostat and light sockets. Take out the center knockout on one side of the box and the center one on a side perpendicular to that one. Fit each of these holes with a ½-inch, 2-screw, clamp-type Romex connector which will fit into the hole. A lock nut will hold it tight from the inside of the box. Leave the clamp part of the connector on the outside of the box. This part will clamp the wires in place once they are installed.

Turn the frame upside down now. You will be mounting the box on the underside of the center brackets. Your box should already have mounting holes, which can be enlarged, if need be, with a drill, using a $^5/_{16}$-inch bit. Be sure that these holes are in line with the center brackets, with one hole on each bracket, and that one Romex connector is aligned with the space between the center brackets. If there are no mounting holes on the box that are in line with the center brackets, drill them, using a $^5/_{16}$-inch bit. Just position the box with the back against the center brackets and one side, with the knockouts intact, even with the inner edge of the side rail, as shown in photo 6-6. Mark center points for the two mounting holes so that each hole will be in the center of the ¾-inch width of the center brackets.

Take the box away to drill the two clearance holes. Then put the box back in place, mark the center points for mounting holes in the center brackets, and drill these two holes, using the same size bit. Now match the holes in the box with the holes in the center bracket, insert a ⅜-inch-long ¼-20 machine screw in each hole, and tighten each with a ¼-20 nut. This will fasten the box to the frame.

The thermostat we recommend will fit between the two center brackets. Take a good look at your thermostat. You'll notice there is a switch with a C-shaped bracket attached to it. A round wafer unit will be in the middle of the C-shaped bracket. When the thermostat is installed on the center brackets, the wafer and the switch will hang below the center brackets with the top part of the C-shaped bracket fastened to the center brackets.

Fit the thermostat bracket in place between the center brackets, keeping the top of the bracket flush with the top of the center brackets. Mark two points on each center bracket near the upper corner of the regulator bracket. Remove the thermostat, and, using a ¼-inch bit, drill a clearance hole through each center bracket on the points marked. Put the thermostat back in place, and mark points for the clearance holes on each side of the thermostat bracket through the holes in the center brackets. Take the thermostat out again, and, holding it firmly against a stable surface, drill clearance holes through each side, using the same size bit. Put the thermostat in place again now, and feed its wires through the connector into the junction box before fastening. Match up the holes and fit them with 1¼-inch-long 10-24 stove bolts, and tighten the bolt with 10-24 nuts. Follow instructions on the thermostat box for adjusting the switch and installing the wafer and adjustment screw.

With all of the electrical fixtures installed on the frame, you are ready to wire the unit. Take a good look at the wiring diagram on blueprint sheet 6 and illustration 6-10. The main thing to remember while doing the wiring is to get the phasing of the wires correct. The positive wire shown black comes into the box and goes to a wire nut. It is joined to one of the thermostat wires, shown gray. The power returns from the thermostat to a wire nut, where it splits to two black wires. These wires go to the lights and are connected to their brass terminals. The negative wire, colored brown, comes into the box and splits at a wire nut and goes out to the two lights' silver terminals. The ground comes in and is fastened to one of the box's mounting bolts. Positive wires are always connected to positive wires and negative to negative. All wire-to-wire connections will be made inside the metal junction box with wire nuts. For added safety the wire nuts should be taped.

All that remains is to attach a three-prong male safety plug, 6H, to the end of the power cord. This will plug into a wall outlet in your home. Take the cover off the plug to expose the terminals. Slide the cover off the plug onto the cord before you connect the wires to the terminals on the other part of the plug. Connect the black wire, the positive, to the brass terminal, and the white wire, the negative, to the silver terminal. The entire frame assembly will fit inside the box, with the end blocks resting on the bottom. Place a 10-inch by 10¼-inch cake-cooling rack on each side of the center brackets. A 10-inch by 10-inch rack may also be used. With this done, you are ready to make tempeh with your homemade tempeh incubator.

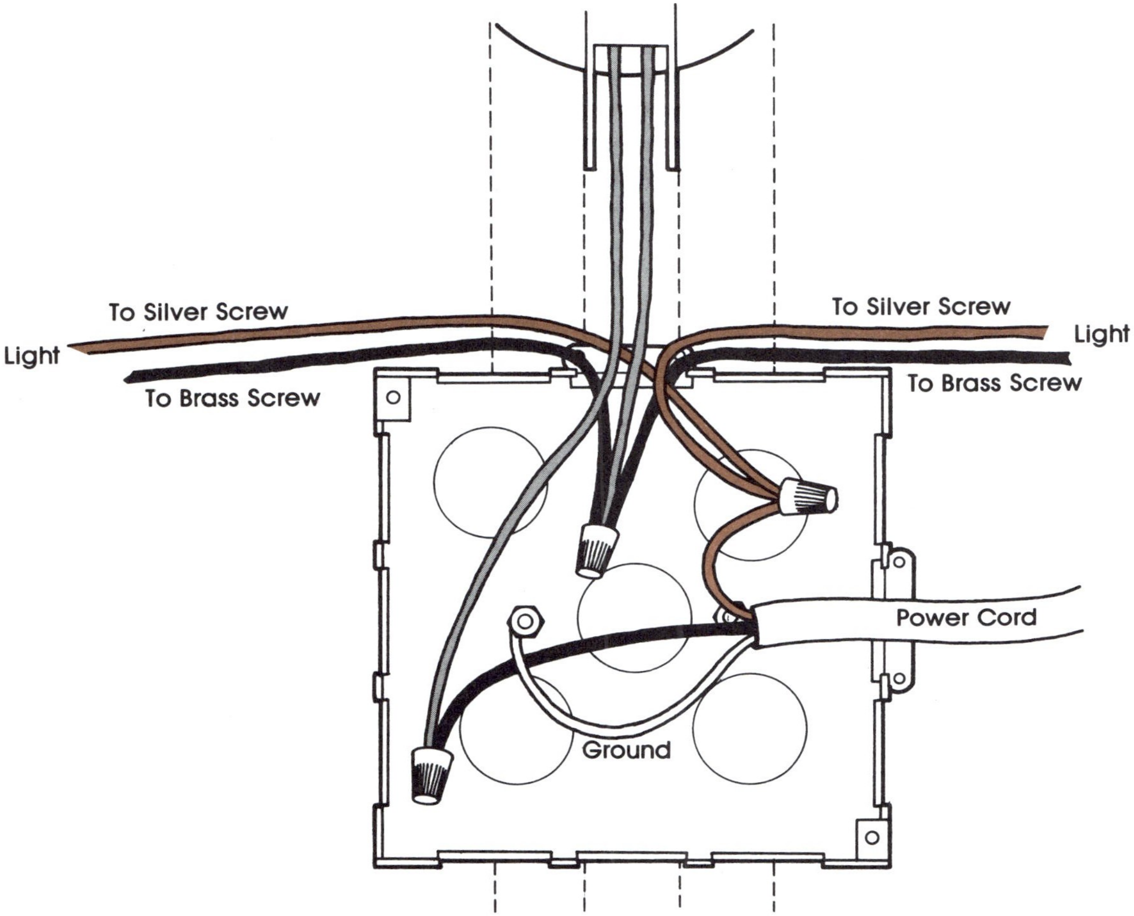

Illustration 6-10—The tray frame is wired as shown. Note that all connections are made inside the junction box.

Section III Blueprints

This section consists of the six blueprint sheets that detail the projects described in section II. As you are building, you will find them invaluable for giving you an overall picture of the project, as well as special detailed illustrations of difficult parts. Each blueprint sheet is perforated so it can be removed from the book. After you are through using them, you can refold them and store them for future reference in the envelope provided.

The format of the sheets is simple. Each sheet is divided into a left and a right section. On the left side, you will see a list of the pieces on that sheet. Reading from left to right, this includes the letter part of the piece's label; the name of the piece as it is used throughout the book; the quantity of that particular piece needed for the assembly; the type of material for the piece; and a description of the piece giving its dimensions, and, in most cases, an illustration of the piece.

Below the chart describing the pieces, there are notes that draw special attention to details that require further explanation. On the right side of the sheet is a series of drawings of the assembly from various views. Each sheet has at least three views of the project.

All of this is to help you to refer to the blueprint sheets and to tie them into what is described in section II. The labels of the pieces as given in section II consist of a number and a letter. The number is the number of the blueprint sheet where you will find that piece detailed. The letter will key the piece to the chart and to the drawings on the right half.

Blueprint Sheet 1—Table-Mount Press. This sheet details the table-mount press, one of the two presses in the book. On the right side of the sheet, the deck assembly and the leg assembly (the two major phases of the project) are shown separately, but they are combined in a perspective view of the completed press at the bottom of the sheet. The plunger and the lever for the press are on blueprint sheet 4.

Blueprint Sheet 2—Pot-Mount Press. The pot-mount press is the other press choice. The pot needed for the pot-mount press is not shown, nor is the pressure system, which includes the lever and the plunger. The pressure system can be found on blueprint sheet 4.

Blueprint Sheet 3—Tofu Box. The tofu box is shown here in three main segments: the sides which form the box, the baffles, and the bottom. They are combined to form the entire completed project in a perspective view at the bottom of the sheet. The pressure plates for the tofu box are included on blueprint sheet 4 as part of the pressure system.

Blueprint Sheet 4—Pressure System. This sheet shows the pressure system, detailing the lever and the plunger assembled in chapters 4 and 5 and the three pressure plates in the chapter 3 assembly.

Blueprint Sheet 5—Tempeh Incubator Box. The tempeh incubator box is detailed on this sheet. Aluminum tape, weather stripping, and rubber feet for the box are not shown in the drawings. Insulation and reflective surface for the interior of the box are not shown here, to give you a better view of the framing of the box.

Blueprint Sheet 6—Tray Frame. The tray frame for the tempeh incubator is shown here without the wiring to avoid confusion. There is, however, a wiring diagram included on the sheet. This sheet, more than any of the others, is not meant to be used without the aid of the text in section II. Follow all safety precautions.

Before you start any assembly, you should look over the blueprint sheets for that assembly to get an overall view of things. In one glance, this will show you just how pieces fit together, giving you a preview of the project and letting you know what to expect. Then as you proceed through the pages in section II, check the blueprint sheet for particular details, and compare the blueprint drawings with the photos and illustrations in the assembly chapter in section II.

If you have any questions on the assembly or operation of the soybean equipment described in this book, contact Reader Service Department, Plans Books, Rodale Press, 33 East Minor Street, Emmaus, PA 18049. Please do not contact us on questions on materials availability. All of the material information that we have is included in this book.